Laboratory Exercises in Invertebrate

By Alan R. Holyo.

First edition copyright © 2013 Alan R. Holyoak

Second edition copyright © 2016 Alan R. Holyoak

All Rights Reserved – text and illustrations except as otherwise indicated

The author retains the right to make copies of the work available for internal distribution by Brigham Young University-Idaho

Dedication

To Todd Newberry, John Pearse, Mike Hadfield, Lee Braithwaite and Larry Hibbert, teachers, mentors and friends who fed my thirst for knowledge about all things invertebrate, and most of all thanks Kathrine for your support and happily joining me on our life-long adventure.

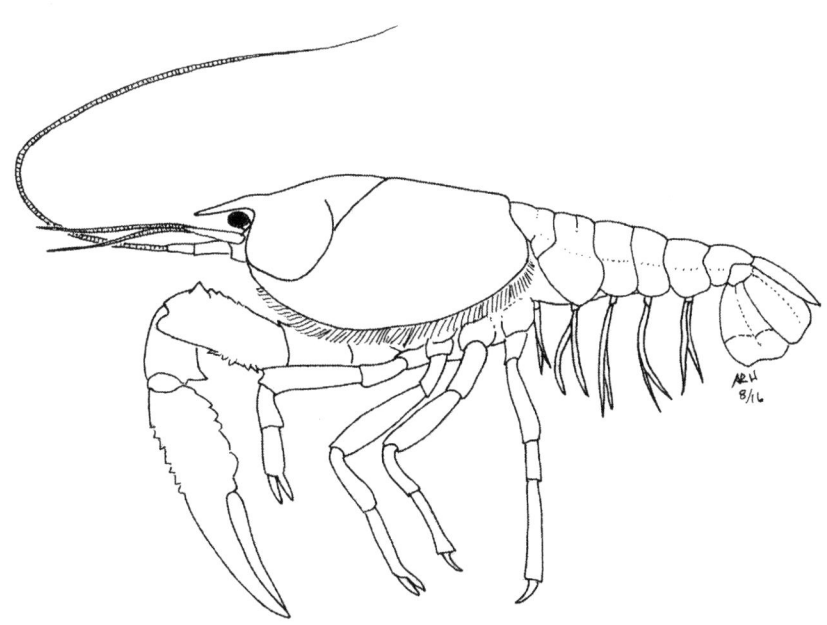

Table of Contents

Preface 5

Chapter 1: Introduction to Microscopy 6

Chapter 2: Phylogenetic Analysis 14

Chapter 3: Domain Eukarya 21

Chapter 4: Phylum Porifera 40

Chapter 5: Phylum Cnidaria and Phylum Ctenophora 50

Chapter 6: Phylum Platyhelminthes 67

Chapter 7: Phylum Mollusca 81

Chapter 8: Phylum Annelida 108

Chapter 9: Phylum Brachiopoda and Phylum Nematoda 121

Chapter 10: Clade Panarthropoda 140

Chapter 11: Clade Deuterostomia, Phylum Echinodermata 178

Chapter 12: Phylum Hemichordata and Phylum Chordata 197

Supplemental Material: The Laboratory Notebook 207

Reference Material 210

Index 212

Preface

Invertebrate zoology is a broad discipline that covers the biology, ecology and evolutionary biology of all animal phyla. Laboratory exercises in invertebrate zoology tend to focus on body plans, life histories and sometimes the behavior of these amazing animals.

This set of exercises is designed to support a one-semester course in invertebrate zoology. It could also support a course in general zoology, at least in part, since these exercises cover taxa from protozoans through invertebrate chordates. Most representatives used in these exercises are available from biological supply companies; this makes these exercises useful for classes and individuals whether they are in coastal or inland locations.

I developed the core of these exercises for my invertebrate zoology course several years ago when I noticed the skyrocketing price of commercially available invertebrate zoology laboratory manuals. In 2013 I published the first edition of these exercises in order to make them more widely available to others who are seeking affordable options to support their own courses in invertebrate zoology. Since that time I identified ways these exercises can and should be improved, including updating taxonomies, adding new figures, improving the supporting narrative text and adding material on protozoans, not to mention finding and correcting those pesky typographical errors that always creep into a project like this. In addition, the publication of *Invertebrates 3e* by Brusca, *et al.* in January 2016 motivated me to update the content of this lab manual to complement that textbook.

The second edition of this laboratory manual is an improvement in many ways over the first edition and I hope it will provide the laboratory support you need. I also hope these lab experiences will help anyone who uses them to become better scientific thinkers and observers. Lastly I am always happy to receive constructive feedback about how these exercises can be improved.

Alan R. Holyoak
Department of Biology
Brigham Young University-Idaho

Chapter 1: Introduction to Microscopy

This laboratory exercise introduces you to principles of microscopy and familiarizes you with compound and dissection microscopes. About now you might be thinking, "Why do I need an introduction to microscopy? I already know how to use a microscope." I don't doubt that many of you have had chances to use microscopes but in my experience too few students receive adequate instruction in microscopy. Since you are now clearly on the path to being scientists it is time that you learned some of the finer points of microscopy.

The compound microscope:

Compound microscopes are used to examine specimens that are small enough, thin enough or transparent enough to allow light to pass through them. These microscopes have a light source in the base that projects light up through a specimen and into the lens system.

Tasks

1) Become familiar with the anatomy of a compound microscope. Use **Fig. 1.1** and related material to do this.

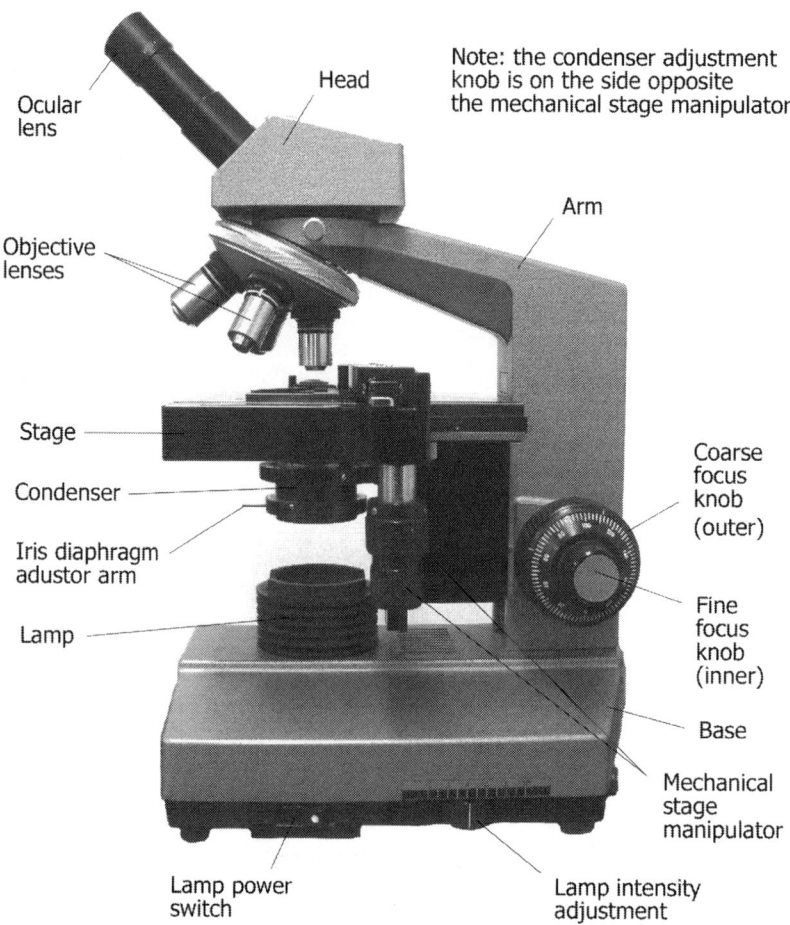

Figure 1.1. A monocular compound microscope. (Image: ARH)

Parts of a compound microscope

    Arm – This structure supports the head of the microscope.

    Base – The base supports the microscope and houses the lamp.

    Coarse/fine focus knobs – The larger outer knob is the coarse focus. Use this knob to find the focal plane for each magnification. The smaller inner knob is the fine focus. Use this knob to fine-tune the focus once you have found the focal plane.

    Condenser – This is located beneath the stage; the condenser focuses light from the lamp to maximize resolution.

    Condenser adjustment knob – This knob moves the condenser up and down to achieve optimal focusing of the light from the lamp.

    Head – The head of the microscope houses mirrors that reflect the incoming light from the objective lens to the ocular lens. The lens system causes the image to be inverted when viewed through the objective lens. The top of the head bears the ocular lens and the underside of the head supports the rotating set of objective lenses. There is a thumbscrew that can be loosened to allow the head to swivel to a preferred position. Re-tighten this screw once the head is in the desired position.

    Iris diaphragm and adjustor arm – The iris diaphragm restricts the amount of light passing from the lamp to the objective lens. The adjustor arm dilates and constricts the iris diaphragm opening. Optimal image quality is obtained by adjusting the lamp intensity and iris diaphragm. Closing the diaphragm increases contrast of the image.

    Lamp – Light from the lamp passes up through the specimen and into the lens system. The notch in the front of the lamp housing is used when you adjust the condenser as described later.

    Lamp intensity adjustment – This sliding lever adjusts the intensity of light produced by the lamp. Less experienced microscopists often set the lamp intensity too high. This washes out the image, fatigues the eyes and can cause headaches. Increase lamp intensity only enough to see the image clearly.

    Lamp power switch – This toggle switches power to the lamp on and off.

    Mechanical stage manipulator – These knobs move the spring-loaded slide holder forward and backward and side-to-side across the stage.

    Ocular lens or eyepiece – This lens slides into the barrel on the top of the microscope head. The magnification of the ocular lens is indicated on the lens and is typically 10x and sometimes 15x.

    Objective lenses – This set of lenses is mounted on a rotating base located on the underside of the microscope head. The magnification power is inscribed on each lens.

    Stage (with spring-loaded slide holder) – The stage supports the slide.

The Dissection Microscope

    Dissection microscopes are used for observing specimens that are too large to observe using a compound microscope or too opaque for light to pass through them. Light reflects off of the surface of these of specimens and up into the microscope.

Tasks

1) Use **Fig. 1.2** and associated material to become familiar with the anatomy of a dissection microscope.

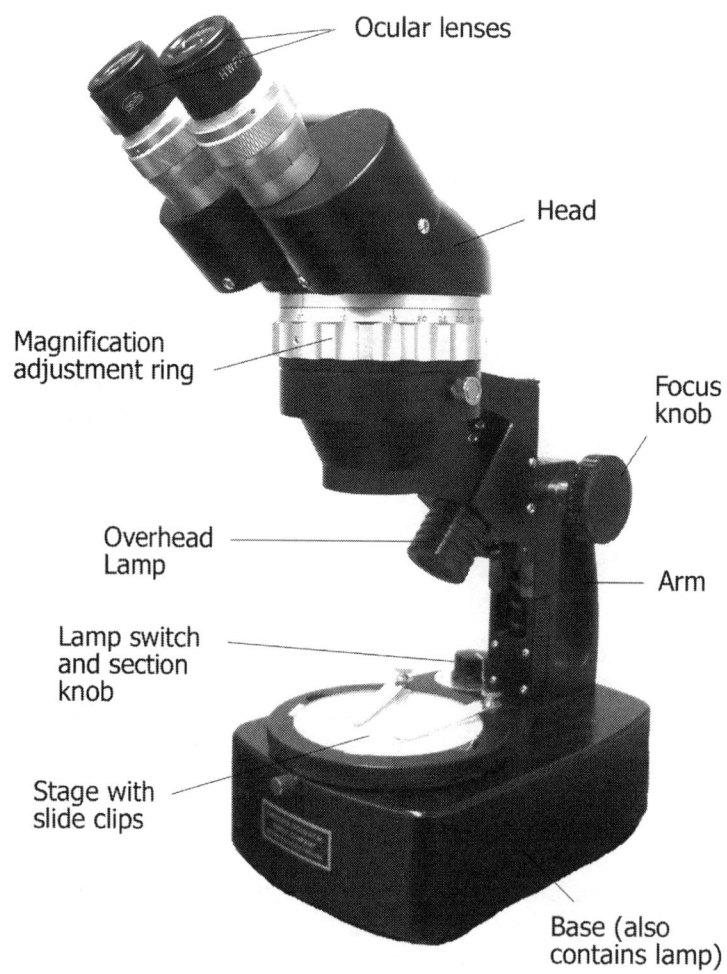

Figure 1.2. A dissection microscope. (Image: ARH)

Parts of a dissection microscope

    Focus knob – This knob is used to adjust focus.

    Head, Arm and Base – These parts fill the same function as described for the compound microscope.

Lamp switch selection knob – This rotating switch allows the user to choose whether to use a built-in overhead lamp, a lamp in the base, both or neither. Light intensity is not adjustable on this particular dissection microscope.

Magnification adjustment ring – Rotation of this ring allows the user to zoom the magnification between 0.7x to 4.5x (multiplied by the power of the ocular lens).

Ocular lenses/eyepieces – These parts have the same function as on the compound scope. Each of the ocular lenses of a binocular scope can usually be focused independently. Use one eyepiece to focus on a specimen, and then focus the other eyepiece.

Overhead lamp – The lamp projects light down onto the specimen.

Stage with stage clips – The stage supports the specimen during observation. Stage clips hold specimens or slides in place. Stage clips can be easily removed. The centerpiece of the stage can be opaque white, opaque black, translucent or clear glass as needed.

Rules of Microscopy

Tasks

1) Learn and follow these basic rules of microscopy
    a. Carrying the scope: ALWAYS use two hands whenever you carry a microscope, one hand goes under the base to support the microscope and the other hand is used to firmly grip the arm of the scope. Cradle the microscope in front of you as you carry it. This keeps the microscope up and away from table corners, doorknobs, etc., and makes it easy for others to see that you are carrying a microscope so they can clear a path for you.
    b. Cleaning lenses: Clean lenses ONLY when they need it. When a lens is dirty use only lens paper to clean lenses – NEVER use any other kind of tissue, paper, Kimwipe™ or cloth to clean a lens – these will scratch microscope lenses.
    c. Focusing on a slide:
        i. To examine a specimen with a compound scope, rotate the lowest power objective lens into position and move the stage to its highest position. Use the coarse focus knob to move the slide stage down and away from the objective lens as you look through the ocular lens and locate the plane of focus. Doing this keeps you from moving a slide up into an objective lens since this could crack or break a slide and possibly damage the lens.
        ii. Once you have found the focal plane using the coarse focus knob, use the fine focus knob to fine-tune the focus. Never rotate the fine focus knob more than one full rotation in either direction as you attempt to focus on a specimen. Use the course focus knob if you have to rotate the fine focus knob more than one full turn without achieving final focus.
        iii. Always work your way up through the objective lenses in order from lowest power to the desired power as you examine a specimen. If you skip a lens you may have a difficult time finding the plane of focus again. If you are using the dissection scope start with the magnification

adjustment ring at 0.7x and increase magnification as needed.
2) Using immersion oil
   a. Use immersion oil if you need to use the 100x objective lens. Immersion oil is used specifically for this purpose. Lenses to be used with immersion oil always have a black ring around the tip of the lens.
   b. Follow this procedure to apply immersion oil.
      i. Find the focal plane for the 100x lens.
      ii. Rotate the objective lenses so the lowest power lens and the 100x oil immersion lens are on either side of the slide.
      iii. Use the built-in applicator in the jar of immersion oil to apply one or two drops of immersion oil to the top of the coverslip of the microscope slide (you should let excess oil drip off of the applicator before moving it to the microscope)
      iv. Rotate the 100x lens into the drop of oil – WARNING – Make sure that you are rotating the 100x lens and not the 40x lens into the oil. Immersion oil will ruin any lens that is not an immersion oil lens. Immersion oil lenses always have a **black ring** around the base of the objective lens, ONLY lenses marked like this should be used with immersion oil.
      v. If you lose the plane of focus, rotate the 100x lens out of position, remove and clean the slide and lens and start over.
3) Procedure for setting up a compound scope before beginning work
   a. Check the microscope
      i. Make sure the lamp works
      ii. Move the stage to its lowest position
      iii. Make sure a slide was not left on the stage by the previous user
      iv. Rotate the lowest power objective lens into place
      v. Check the lenses to make sure they are clean; clean them ONLY if needed.
   b. Adjust the condenser
      i. Use the lowest power objective lens to focus on a slide.
      ii. Use the condenser adjustment knob to move the condenser to its highest position
      iii. Place the tip of a probe so that it rests horizontally on the notch at the front of the lamp housing and its tip rests lightly but directly touching the center of the lamp.
      iv. Look through the ocular lens while holding the probe tip in position and rotate the condenser downward using the condenser adjustment knob. Continue to lower the condenser until you see a sharp outline of the silhouette of the probe tip in the field of view.
      v. When the silhouette of the probe tip is in sharp focus you may also see a mottled pattern in the rest of the field of view. This mottling is produced by imperfections in the glass surface of the lamp. Use the condenser adjustment knob to move the condenser either slightly up or down, just enough so that the mottling disappears. Your condenser is now correctly adjusted and you can get to work.
1) Procedure for preparing a microscope when you are done using it
   a. Remove the slide from the stage and put the slide away

b. Use a paper towel damp with tap water to remove any salt water remaining on the microscope and wipe the scope with a dry paper towel to remove any other water on the microscope
   c. Use the coarse focus knob to move the stage to its lowest position
   d. Turn the lamp off
   e. Wrap the cord loosely around the base (wrapping it too tightly can damage the cord).
   f. Replace the dust cover if one is provided.

Make a wet-mount slide

You occasionally need to make your own slides of specimens. To make a temporary wet-mount slide one you need the following materials: a specimen, a microscope slide, a cover slip, a pipette or eyedropper and some plasticene clay.

Task

1) Make a wet-mount slide by following this procedure:
   a. Place your specimen in the middle of a glass slide and use a pipette to place a couple of drops of water on top of the specimen. Make sure that there are no air bubbles clinging to the specimen.
   b. Prepare the cover slip by holding some plasticene clay in one hand and a cover slip in the other. Gently drag each corner of the cover slip across the surface of the clay. A small amount of clay should now be on each corner of the cover slip. These small bits of clay serve as a tiny support posts to keep the weight of the cover slip from flattening the specimen.
   c. Lay one edge of the cover slip down so it extends from one side of the microscope slide to the other with the plasticene clay on the underside of the coverslip.
   d. Support the opposite edge of the coverslip with a probe tip. Lower the coverslip slowly until it comes in contact with the drop of water covering the specimen. Capillary action will draw the water out between the coverslip and the glass slide as you lower the coverslip all the way down. Lowering the cover slip slowly minimizes the number of air bubbles that get trapped under the coverslip.
   e. If the cover slip floats there is too much water; use the corner of a paper towel to touch the space between an edge of the coverslip and the glass slide. The coverslip will settle onto the glass slide as water wicks out.
   f. Water will not reach the edges of the coverslip if there is too little water. If this is the case use a pipette to add more water to the space between the coverslip and the slide.
   g. Your slide is now ready for observation. Keep in mind that the coverslip is not physically attached to the glass slide so you need to handle the slide gently. When you are done with your wet-mount slide your instructor should tell you what to do with the cover slip and slide.

Generate a scale bar for a lab drawing

Tasks

1) Produce a scale bar for a drawing of a specimen that is large enough to measure with a ruler:
    a. Produce a drawing in your laboratory notebook
    b. Measure the maximum length of your drawing
    c. Measure the same dimension on your specimen
    d. Determine what scale you want to represent, make it a nice standard number like 1.0 mm, 1.0 cm, 5.0 cm, etc.
    e. Multiply the length of your drawing by the scale you desire divided by the length of the specimen.
    f. The answer is the length of line you need to draw to represent the desired scale.

    Example:
    The length of your drawing = 12.5 cm
    The actual length of the object = 4.3 cm
    You want the scale bar to represent 1.0 cm
    Scale bar length = 12.5 cm x (1.0 cm / 4.3 cm) = 2.9 cm

    In this example you would draw a line 2.9 cm long someplace near the drawing in your laboratory manual and write "1.0 cm" above that line. This means that 2.9 cm on the drawing represents 1.0 cm on the original specimen, or that your drawing is nearly three times as large as the specimen.

2) Produce a scale bar for a specimen observed via microscope. This how to do this when your microscope does not have an ocular micrometer:
    a. Use a stage micrometer to measure the width of the field of view for each magnification of your compound microscope. Record these field width measurements in the front of your lab notebook where you can easily find them; you will refer to these field width measurements throughout the course.
    b. Draw the specimen you are observing.
    c. Measure the length of your drawing.
    d. Estimate the length of your specimen using the field widths you recorded earlier.
    e. Calculate the length of the scale bar by multiplying the length of your drawing by scale you desire divided by the actual length of the specimen.

    Example:

    The length of your drawing is 8.7 cm
    The actual length of the object you drew is 2575 μm
    You want to produce a scale bar that represents 100 μm on the original object

In this example the scale bar length = 8.7cm x (100μm/2575 μm) = 0.34 cm. Oops! This length (0.34 cm) is too short to draw with much precision so in this case you should produce a scale bar that represents a greater length, say 500 μm. To do this multiply your answer (0.34 cm) by five and then draw a line 1.7 cm long and write "500 μm" over the line. Done!

3) Produce a scale bar for a specimen observed via microscope. This how to do this when your microscope has an ocular micrometer:
    a. Calibrate your ocular micrometer for each objective lens as follows:
        i. Focus on a stage micrometer with the lowest power objective lens
        ii. Rotate the ocular lens bearing the ocular micrometer so lines on the ocular micrometer are parallel to stage micrometer lines.
        iii. Move the stage micrometer so the stage and ocular micrometer lines partially overlap each other (one above the other) and the far left line on the ocular micrometer and the far left line on the stage micrometer line up.
        iv. Scan the ocular micrometer starting at the right end and look for the first place an ocular micrometer line and stage micrometer line up.
        v. Divisions of the stage micrometer are 10 μm apart, so divide the number of ocular micrometer units between the left end and the marks that line up at other end by the distance between those marks on the stage micrometer. This tells you the width of an ocular micrometer division for the lowest power objective lens.
        vi. WRITE this number in the front of you lab notebook next to the objective lens magnification and repeat this procedure for all other objective lenses on your microscope.
    b. Draw a specimen you are observing under the microscope.
    c. Measure the length of your drawing.
    d. Measure the length of your specimen using your calibrated ocular micrometer and then calculate the length of the scale bar as described above.

You are now prepared to use microscopes in your investigations of invertebrates.

Group Questions
1) What are advantages and limitations of using a compound microscope?
2) What are advantages and limitations of using a dissection microscope?
3) What is the benefit of producing a scale bar for each drawing you make?

Chapter 2: Phylogenetic Analysis

Scientists use phylogenetic analysis, also known as cladistics to test hypotheses about relatedness between taxa. You will see many cladistic trees as you study zoology. Since this is the case it is a good idea to know something about how they are made.

The goal of this exercise is to introduce you to the basics of cladistics and to give you a chance to work through an exercise where you develop a phylogenetic tree. Before you can do this however you need to be familiar with the vocabulary of this discipline.

Task #1: Review the following terms of cladistics.

- **Taxon** – Any scientifically named group of organisms; a taxon can be as small as a species or as large as a domain.
- **Ancestral taxon** – A taxon that gave rise to at least one descendant taxon by speciation.
- **Daughter taxon** – A taxon that descended from an ancestral taxon by speciation.
- **Monophyletic group or Clade** – A group of taxa that includes an ancestral taxon and all of its descendant taxa. The identification of monophyletic taxa is the goal of cladistics.
- **Paraphyletic taxon** – A violation of monophyly because at least one descendant taxon of an ancestral taxon is not included in the grouping.
- **Polyphyletic taxon** – A violation of monophyly because more than one ancestral taxon is required to describe the origins of the taxa within the group.
- **Ingroup** – The group of taxa you are investigating.
- **Sister taxon** – The most closely related taxon to the ingroup that is not part of the ingroup.
- **Outgroup** – A taxon, ideally the sister taxon, that is included in phylogenetic analysis for comparative purposes to help identify evolutionary polarity of traits.
- **Character** – An anatomical or genetic trait of a taxon. Be advised that a character can be gained and then lost during evolution but the probability of that exact same character being regained is so low that we do not consider it to be a possibility during analysis.
- **Symplesiomorphy** – A shared ancestral trait, i.e., a trait a taxon inherited from its ancestor. Do not use symplesiomorphic characters to do cladistics.
- **Synapomorphy** – A shared derived trait, i.e., a trait a descendent taxon has that its ancestor did not have. Use synapomorphic characters to do cladistics.
- **Character polarity** – An informed decision about the relationship between different versions or states of a particular character. The outgroup is used to determine which version of a character state is ancestral and which is derived.

The goal of cladistics is to describe the evolutionary relatedness between taxa and to resolve problems of paraphyletic and polyphyletic taxa so that all taxa are monophyletic. Cladistics includes the practice of producing cladograms, also known as phylogenetic trees based on parsimony. Though not definitively correct, the most parsimonious tree is the one that has the fewest number of character changes while still preserving the existence of identifiable synapomorphies and unless there are compelling reasons to the contrary these are generally accepted as being the most likely.

Professional taxonomists do not do phylogenetic analysis by hand, instead they computer programs to generate cladograms. In this exercise however you will produce a cladogram by hand to give you insight into how phylogenetic analysis is done.

Procedure for constructing a cladogram

1. Identify the ingroup, outgroup and synapomorphic characters to use in the analysis.
2. Produce a matrix of characters and taxa
    a. Write the names of the outgroup taxon and ingroup taxa along the top of the matrix.
    b. Write the names of characters down the side of the matrix.
3. Find out which characters each taxon has and enter an appropriate notation in each cell for each taxon and character combination.
    a. This is relatively easy as long as a trait has only two options: present or absent. If a character is present in a taxon write "1" in the cell. If it is absent write "0" in the cell.
    b. Things get a bit trickier when there are multiple states for a character. For example, these are four possible options for body symmetry: asymmetrical, radial, biradial and bilateral. We assign these the numbers 1, 2, 3 and 4 respectively for the character. These numbers represent the character polarity from ancestral to derived character states.
    c. Once you have filled all cells in the matrix you are ready to begin the analysis and develop your cladogram.
4. Your instructor will direct you to either choose your own ingroup and outgroup and develop a matrix for them or use the ingroup, outgroup and characters provided in this exercise.
5. A cladogram is complete when all taxa and character combinations are accounted for on the cladogram. A cladogram is basically a branching tree that includes all taxa and all characters in the matrix.
6. The cladogram has a long straight line called the backbone and side branches that come off the backbone. Characters are indicated either on the backbone or on the side branches. Taxa are found only at tips of the branches. The easiest way to explain how to do cladistic analysis is via an example.

Task #2: Work through this example of how to generate a simple cladogram

Taxa and characters:
Outgroup: Prokaryotes (I lumped all prokaryotes together for convenience.)

Ingroup: Kingdoms Animalia, Fungi, Plantae and Protista (Protista is no longer a viable taxon but I use it here for convenience.)

Characters: 1) DNA, 2) Nucleus, 3) Chloroplasts (present in at least some members of the taxon), 4) *Hox* genes, 5) Cell walls (not peptidoglycan) and 6) Strict multicellularity (all species in a taxon must be multicellular).

This is a list of the taxa and their characters:
- **Prokaryotes**: DNA
- **Animalia**: DNA, Nucleus, *Hox* genes, Strict multicellularity
- **Fungi**: DNA, Nucleus, Cell Wall, Strict multicellularity
- **Plantae**: DNA, Nucleus, Chloroplasts, Cell Walls, Strict multicellularity
- **Protista**: DNA, Nucleus, Chloroplasts, Cell Wall

1) Draw a matrix, i.e., a table with a column for each taxon and a row for each character. Write a "1" in each cell of the table where a taxon has a particular trait and write a "0" in the cell where it does not. Once your matrix of taxa and characters is complete you are ready to generate a cladogram.
2) Draw a long diagonal line on your paper. This is the backbone of the cladogram.
    a. A character that appears on the backbone is found in all taxa farther up the tree.
    b. Any character that appears on a side branch applies only to the taxon or taxa on that side branch.
    c. Start by drawing a short hash mark across the backbone near its base and write "DNA" next to it. All taxa have this trait. This indicates that all taxa on the tree have this trait as indicated in your matrix by the complete row of ones next to the character DNA.
3) By definition the outgroup branches off first. If you look at your completed matrix you will see that Prokaryotes have only DNA. Draw a line extending perpendicularly to the backbone but above the hash mark for DNA and write "Prokaryotes" at its tip. The taxon Prokaryotes and all of its characters are now accounted for.
4) Decide which taxon of the ingroup branches off first. How can you decide which of the ingroup taxa should branch off first? Look at your matrix and you will see that three of the four ingroup taxa have four characters in their columns while Plantae has five. You now have to make a judgment call. Since this may be your first try at cladistics call it a hypothesis and choose Animalia, Fungi or Protista to branch off next. Feel free to use prior knowledge and common sense as you do this – that's what brains are for. For the purposes of this example I decided to have Protista branch off next. Protista has three traits that Prokaryotes did not: Nucleus, Chloroplasts and Cell Walls.
5) Decide which of these characters should go on the backbone between Prokaryotes and Protista and which should go on the side branch to Protista. Nucleus is easy. It goes on the backbone because all remaining taxa have this trait. The other two traits are tougher because all remaining taxa do not have Chloroplasts and Cell Walls. You can handle this in different ways; two of these are shown on the cladograms in **Figs. 2.1 & 2.2**.
6) Continue adding characters and branching off taxa until all taxa and characters are accounted for on the cladogram.
7) The cladograms in **Figs. 2.1 & 2.2** were both produced using the same data from the same matrix. Check to make sure that all taxon and trait combinations are present on both cladograms. Which cladogram is correct? Is either correct? The first question you should ask is, "Are they equally parsimonious?" They both have eight character changes, i.e., places where characters appear or disappear so they are equally parsimonious. Even so the cladograms show fundamentally different evolutionary scenarios.
8) Take time with a partner or in a small group to review each cladogram while referring to your matrix. Compare and contrast the two cladograms and think about why the

characters were placed where they were. Discuss these differences between the cladograms. WRITE your observations in your lab notebook.

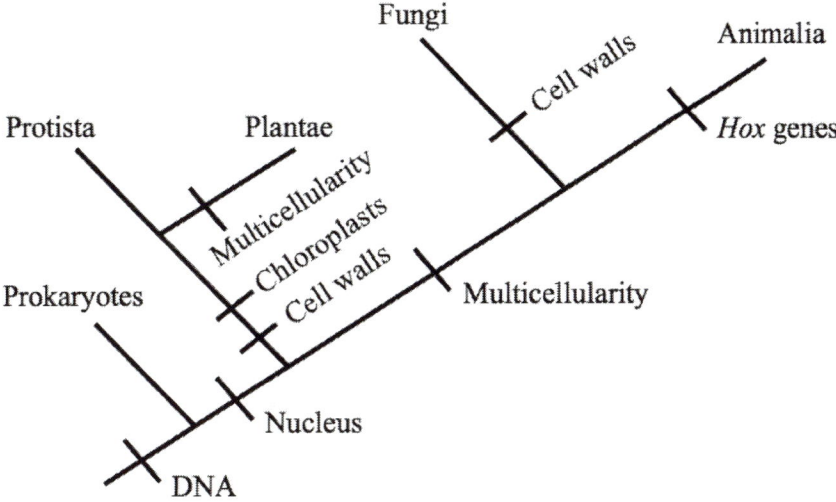

Figure 2.1. Cladogram #1, developed from the data provided above. (Image: ARH)

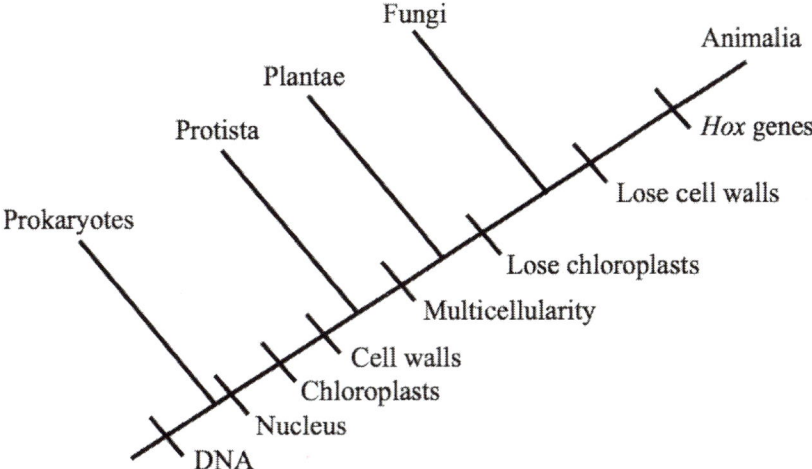

Figure 2.2. Cladogram #2, developed from the data provided above. (Image: ARH)

Task #3: Produce a cladogram

Use the data below to develop your own cladogram. Don't worry if you have never done this before, most of your classmates will be in the same situation so don't hesitate to ask questions or share your thoughts and ideas with each other.

Characters to include in the analysis:

1. Cadherins (cell adhesion molecules): No=0, Yes=1.1
2. Multicellular body: No=0, Yes=2.1
3. Body symmetry
   3.1. Asymmetrical
   3.2. Radial
   3.3. Bilateral
4. Cnidocytes: No=0, Yes=4.1
5. Molted external cuticle: No=0, Yes=5.1
6. Segmented body: No=0, Yes=6.1
7. Shell-secreting mantle: No=0, Yes=7.1
8. Water vascular system: No=0, Yes=8.1
9. Dorsal Hollow Nerve Cord: No=0, Yes=9.1

Taxa to include in the analysis:

Outgroup taxon (characters from the list above are indicated)

- Choanoflagellata: 1.1

Ingroup taxa - listed alphabetically and characters from the list above are indicated for each taxon.

- Annelida: 1.1, 2.1, 3.3, 6.1
- Arthropoda: 1.1, 2.1, 3.3, 5.1, 6.1
- Chordata: 1.1, 2.1, 3.3, 6.1, 9.1
- Cnidaria: 1.1, 2.1, 3.2, 4.1
- Echinodermata: 1.1, 2.1, 3.3, 8.1
- Mollusca: 1.1, 2.1, 3.3, 7.1,
- Nematoda: 1.1, 2.1, 3.3, 5.1
- Platyhelminthes: 1.1, 2.1, 3.3
- Porifera: 1.1, 2.1, 3.1

1) Let's do the first few steps together. Create a matrix using the lists of taxa and characters listed above (see **Table 2.1**). Fill in each cell of the matrix indicating which taxa have which characters. Put a "0" in every cell where a taxon lacks a character.
2) Draw a backbone just like you did for the previous example.
3) Choanoflagellata is the Outgroup so by definition it is the first taxon to branch off of the backbone. Look in the column in your matrix below Choanoflagellata and you will see that it has only character 1.1. Make a hash mark near the bottom of the backbone of the cladogram and write "1.1" next to it. Next, draw a line perpendicular to the backbone and just above that hash mark and write "Choanoflagellata" at the end of that line. You are now done with the Choanoflagellata.

Table 2.1. Matrix of characteristics and taxa - the first two characters are entered, all cells need to be filled in before a tree can be generated.

Characters	Choanoflagellata (outgroup)	Annelida	Arthropoda	Chordata	Cnidaria	Echinodermata	Mollusca	Nematoda	Platyhelminthes	Porifera
1	1.1	1.1	1.1	1.1	1.1	1.1	1.1	1.1	1.1	1.1
2	0	2.1	2.1	2.1	2.1	2.1	2.1	2.1	2.1	2.1
3										
4										
5										
6										
7										
8										
9										

4) Arrange the taxa in ascending order of number and evolutionary state of characters. Ingroup taxa are listed alphabetically in the matrix (**Table 2.1**), so they are almost certainly not in the order they will appear on the completed cladogram. One way to move forward from this point is to make a copy of the completed matrix and cut it into vertical strips with each taxon having its own strip. Arrange the strips so that taxa with the fewest characters and the lower numbered character states are farthest to the left (closest to the Outgroup) and taxa with more characters and larger character state numbers are farthest to the right. When you do this you should see that Porifera is a good candidate to line up next to the Choanoflagellata. Porifera has only three characters and it shares one of them with Choanoflagellata. The shared character is already written on the backbone of your cladogram so you don't need to write it again for Porifera. Porifera however also has characters 2.1 (Multicellular) and 3.1 (Asymmetrical body). Make another hash mark on the backbone just above the line going to Choanoflagellata and write 2.1 next to it since Porifera and all other remaining taxa have character 2.1. What about character 3.1, Asymmetrical body? Should this character go on the backbone or on the side branch going to Porifera? Discuss this question with a partner or in a small group and then place character 3.1 on the cladogram where you think it is most

appropriate. Draw another line perpendicular to the backbone above the hash mark for 2.1 and write "Porifera" at the end of it. You are now done with Porifera in the analysis.

5) Whenever you enter a higher character state for a particular character on the backbone or a side branch it replaces the earlier ancestral character state. For example if you put character 3.1 on the backbone it will be replaced by character 3.2 or even 3.3 if 3.2 is located only on a side branch. The assumption is that all taxa farther up the backbone will have whatever the last character state was until it is replaced by a more derived character state.

6) Repeatedly check your cladogram as you add taxa and characters to see if a trait is more appropriate on the backbone or on a side branch. For example, taxonomists once thought that segmented bodies evolved only once and that all animals with segmented bodies were closely related to each other. Recent research strongly suggests however that segmentation evolved independently multiple times. You will have to decide how to apply this information as you develop your cladogram.

7) Continue to add characters and taxa until all taxa and character combinations from the matrix are accounted for. Keep in mind that it is possible for a trait to evolve and then disappear, but once this happens that same trait cannot reappear on your cladogram. You can indicate the loss of a character trait by writing the trait with a minus sign in front of it, e.g., -2.1. However, you do not do write a character with a minus sign in front of it on your cladogram when it is replaced by a more derived character state.

8) Your goal is to produce the most parsimonious cladogram possible, i.e. the one with the fewest character changes (additions or losses of traits). Chat with each other as well as with other groups as you add more taxa and characters to your cladogram. Science is after all a discipline of developing hypotheses, i.e., possible explanations and communicating conclusions. Your cladogram will be complete once all taxa are present and every character for every taxon is accounted for.

9) INCLUDE your final matrix and cladogram in your laboratory notebook along with observations and questions you have about generating cladograms.

Another approach you could take to developing your cladogram is to draw a tree complete with all taxa but no characters indicating where you think they should be in relation to each other evolutionarily, and then add all of the characters where they fit onto your cladogram. If you do this exercise this way the tree you start with represents your hypothesis and the total number of resulting character changes it takes to make it work indicates its degree of parsimony.

Group Questions

1. In what ways are the cladograms in Figures 2.1 and 2.2 alike? In what ways are they different?
2. Reflect on the cladograms in Figures 2.1 and 2.2 and describe the different evolutionary scenarios they present.
3. Comment on any unexpected results that appeared in your final cladogram of the invertebrates.

Chapter 3: Domain Eukarya

Our taxonomic scheme of life remained largely unchanged from the time of Linnaeus and Cuvier in the late 1700s through the mid-twentieth century. Their taxonomy had two kingdoms: Animalia and Plantae. In 1969 the Five-Kingdom Model of life was proposed. It contained Kingdoms Bacteria, Protista (Protoctista), Fungi, Plantae and Animalia. Then in 1977 the application of molecular biology to phylogenetic analysis led to the development of the current Three-Domain Model that contains the superkingdoms Bacteria, Archaebacteria and Eukarya.

In the Three Domain Model the old eukaryotic kingdoms Animalia, Plantae and Fungi remained largely intact as monophyletic clades within Superkingdom Eukarya. Kingdom Protista (Protoctista) was however identified as being woefully polyphyletic and in dire need of taxonomic revision. This is no surprise since Kingdom Protista was for many years the eukaryotic garbage can kingdom; the place where eukaryotic taxa were dumped that didn't fit easily into any of the other kingdoms.

Work on the taxonomy of Superkingdom/Domain Eukarya "true nucleus" stemming mainly from much needed revision of the now defunct Kingdom Protista resulted in a vastly improved taxonomy for all eukaryotes. Though a new taxonomy of eukaryotes is by no means final, one recently proposed taxonomic model (one of many that are out there) contains five eukaryotic super-groups: Amoebozoa, Chromaveolata, Rhizaria, Excavata and Opisthokonta (see **Table 3.1**). One thing I can say about work on the taxonomy of eukaryotes is that you should be surprised only if there are no further changes. Though Kingdom Protista is no longer accepted as a clade the terms protist and protozoan are still used widely as terms of convenience.

Table 3.1. One proposed taxonomy of Domain Eukarya (after Brusca, *et al.* 2016).

Groups	Selected characteristics	Selected representatives
Amoebozoa	Mostly free-living and unicellular with broad rounded pseudopodia, strictly heterotrophic, some store carbohydrates	amoebae, cellular slime molds, plasmodial slime molds
Chromalveolata	Mostly free-living and unicellular, heterotrophic, autotrophic, mixotrophic or parasitic. Plastids are the result of secondary endosymbiosis with a unicellular red alga (lost in some taxa), flattened membrane-bound alveoli common in many	dinoflagellates, apicomplexans, ciliates, brown algae, coccolithophores, cryptomonads
Rhizaria	Free-living, mixotrophic, parasitic and heterotrophic forms, amoebae with long, thread-like filopodia, some secrete $CaCO_3$ or SiO_2 tests	foraminiferans and radiolarians
Excavata	Mostly heterotrophic with anterior flagellae, some mixotrophic and parasites, many with a ventral feeding groove, taxonomy remains problematic	*Trichomonas, Giardia, Euglena, Trypanosoma*
Opisthokonta	Unicellular to multicellular, flagellate cells bear one posterior flagellum (lost in some), common biochemical traits	choanoflagellates, animals, fungi, red and green algae and all plants

Protists are covered in this lab manual since this may be the only opportunity many of you will have to be introduced to these groups and because a group of protozoans is the sister taxon to metazoans. Selected representatives of the five proposed eukaryotic super-groups are represented.

Task #1 - Group 1 Amoebozoa (word root = "amoeba animals")

Phylum Amoebozoa

1) Observe and describe the anatomy and behavior of any available live specimens. Describe and DRAW what you see.
2) Observe mounted and stained specimens of the following species. Describe and DRAW what you see.
   a. *Amoeba proteus* (*Chaos diffluens*) - Free-living heterotrophic species in freshwater habitats. Prey includes other protozoans, algae and micro-metazoans. This is a commonly used representative of amoeboid protozoans. Note the broad pseudopodia also called lobopodia. Color of living specimens may vary in relation to phytochromes they possess. The clear zone just within the end of a pseudopod is the ectoplasm. The ectoplasm lacks organelles and other structures found in the endoplasm. Use **Fig. 3.1** to help you identify what you see.

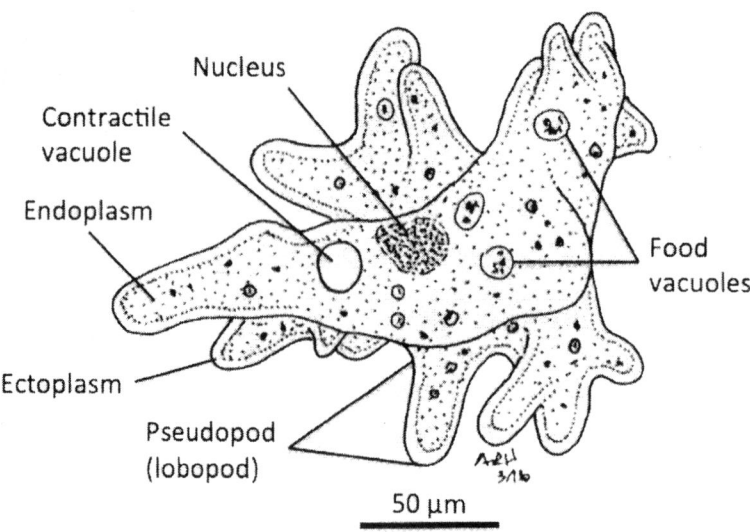

Figure 3.1. *Amoeba proteus* (*Chaos diffluens*). (Image: ARH)

   b. *Pelomyxa* (*Chaos*) *carolinensis* – Macroscopic (1-5 mm) heterotrophic scavengers and predators of everything from bacteria through micro-metazoans in freshwater habitats. This species may have as many as 1000 nuclei. Use **Fig. 3.2** to help you identify what you see.
   c. *Entamoeba* (*Endamoeba*) *hystolytica* – Anaerobic amoeboid found in humans and other primates. Look at **Fig. 3.3** to see cells in place in the gut and **Fig. 3.4** to see a life cycle of this parasite.

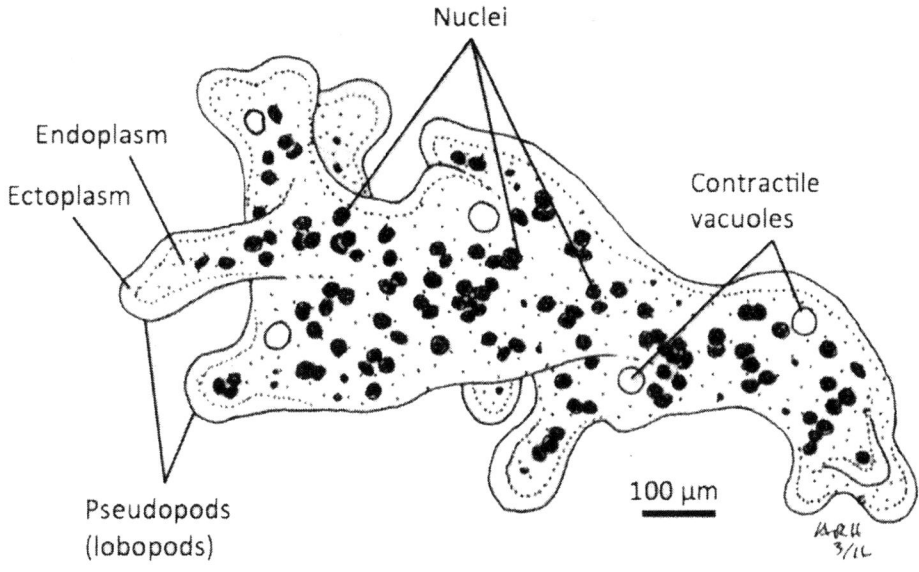

Figure 3.2. *Pelomyxa carolinensis* (Image: ARH)

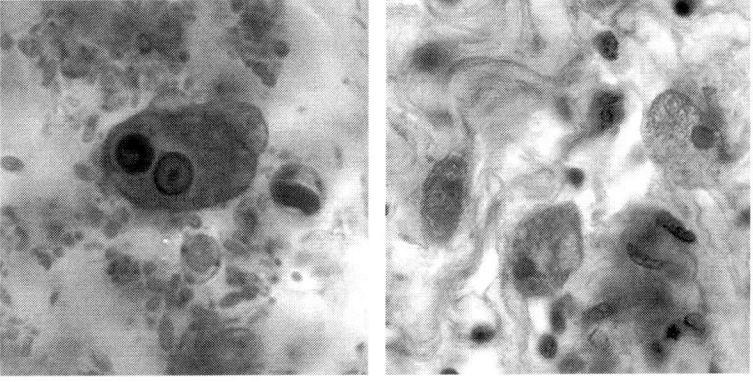

Figure 3.3. *Entamoeba* (*Endamoeba*) *hystolytica*. **Left** - trophozoite in smear preparation; **right** – *E. histolytica* cells in colon tissue, dark inclusions in both images are RBCs. (Image: ARH modified CDC images, http://www.cdc.gov/dpdx/amebiasis/index.html).

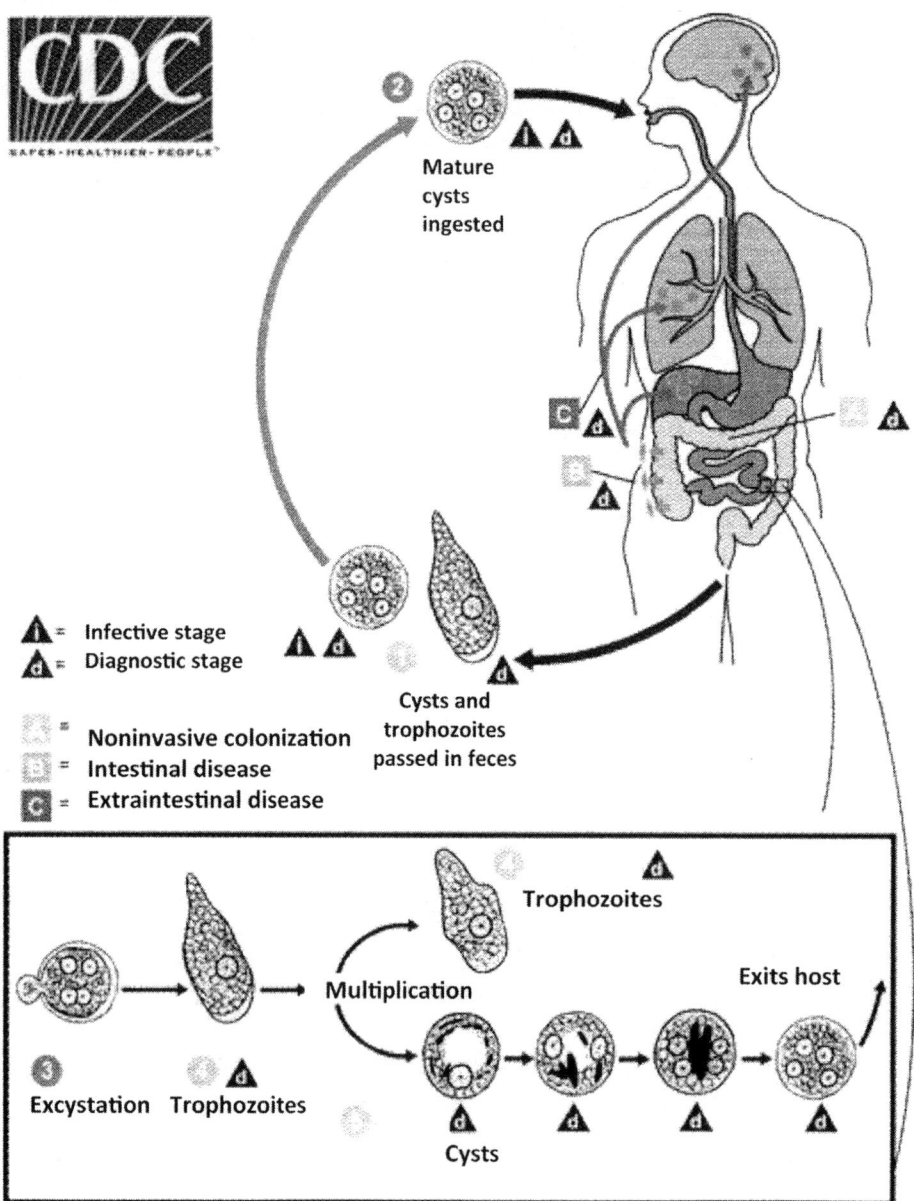

Figure 3.4. Life cycle of *Entamoeba hystolitica*. Cysts and trophozoites are released in the feces (1). Cysts can persist in the environment for up to weeks but trophozoites die soon after release. Infection can occur via ingestion of food or water or hands contaminated with fecal material bearing cysts or during sexual contact (2). Excytsation occurs in the small intestine and trophozoites move to the large intestine where they undergo binary fission and take up residence in the intestine wall (3-5). Infection is asymptomatic when these parasites stay in the intestine. Cells that enter the bloodstream can however affect the liver, brain or lungs. According to the CDC cells that invade tissues other than the lining of the large intestine are a different species, *Entamoeba dispar*. Trophozoites in the lining of the intestine produce cysts and other trophozoites. Cysts are released with formed feces and trophozoites are typically released with diarrhea. (Image: ARH modified image from CDC, http://www.cdc.gov/dpdx/amebiasis/)

Task #2 Group 2 Chromalveolata "colored cavity or pit"

Phylum Dinoflagellata "whirling flagellate"

1) Observe, describe and DRAW any available dinoflagellates. Dinoflagellates have two flagellae. The transverse flagellum runs around the middle of the cell in a groove called the girdle and is attached to the cell wall via a thin transparent and membranous structure. The origin of the longitudinal flagellum is housed in a groove called the sulcus, and this flagellum extends freely into the environment. Flagellae are delicate structures that may not be visible or even preserved when specimens are stained and mounted permanently on slides. An outer protective layer of cellulose plates protects the body. Some species of dinoflagellates are notorious for producing red tides that cause shellfish poisoning in people who eat clams that have been feeding on these protists. Use **Fig. 3.5** to help you identify what you see.

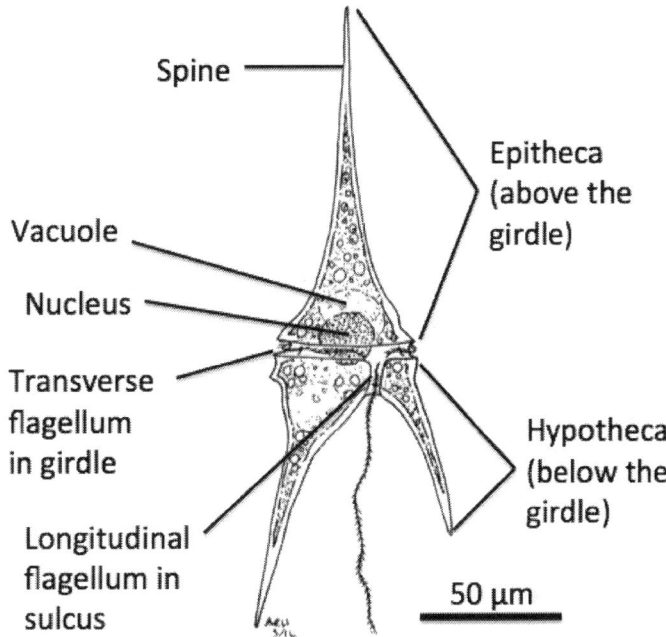

Figure 3.5. The dinoflagellate *Ceratium sp.* (Image: ARH)

Phylum Apicomplexa "tip complex"

1) Observe and describe any available specimens representing this group and describe and DRAW what you see.
    a. *Eimeria sp.* – Parasitic species that causes a disease called coccidiosis in cattle and chickens. See **Fig. 3.6** to see cells of this species in the lining of the large intestine. Animals are infected with *Eimeria* when they ingest oocysts as they ingest material contaminated with feces. Excystation occurs in the small intestine and sporozoites are released into the lumen of the intestine. These burrow into cells lining the intestine where they undergo clonal growth producing first-

generation merozoites. These reinfect cells of the small intestine wall and produce a second generation of merozoites. These make their way to the large intestine where they take up residence in cells lining the large intestine and become gametocytes. Microgametes fertilize macrogametes that develop into oocysts that are released in the feces. Bloody diarrhea results when spores are released across walls of the large intestine and this can be fatal in cases of infection with extremely high numbers of parasites, but is asymptomatic in most cases.

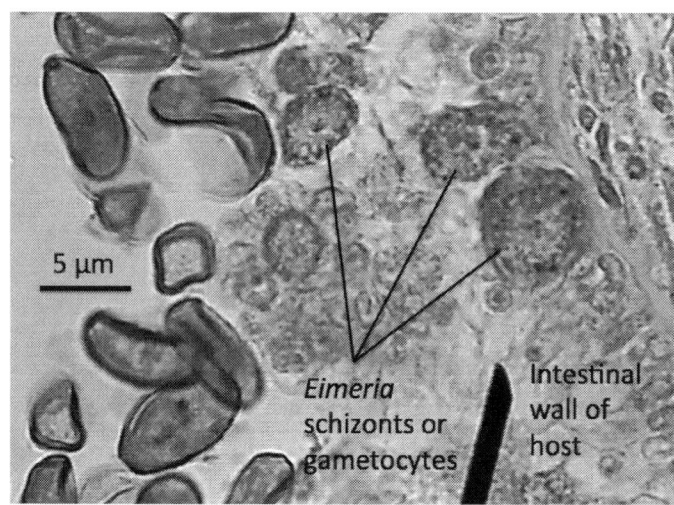

Figure 3.6. *Eimeria sp., in situ.* This parasite causes coccidiosis. (Image: ARH)

b. *Plasmodium sp.* – Members of this genus cause malaria. The WHO reports that over 3 billion people are at risk from malaria, and that there were over 200 million reported cases worldwide in 2015 and over 400,000 deaths. The pathogen is transmitted via an insect vector; female *Anopholes* mosquitos. This makes malaria one of the greatest health risks on earth. **Figure. 3.6** shows malarial trophozoites of this pathogen. **Figure 3.7** shows the life cycle of *Plasmodium*.

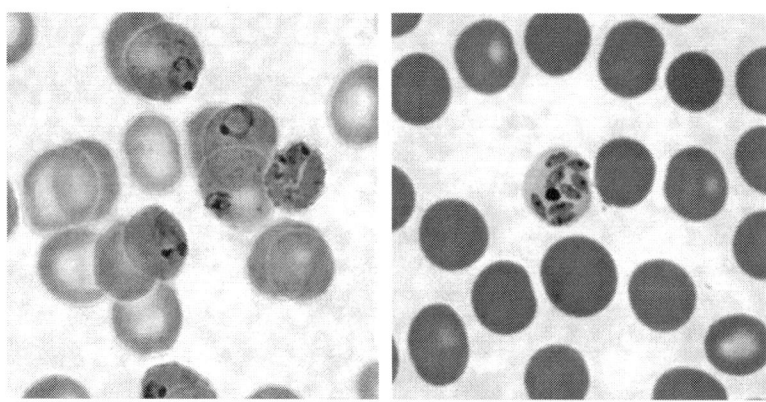

Figure 3.6. *Plasmodium sp.*, trophozoites in RBCs in blood smear, **left**; schizonts in RBC in blood smear, **right**. (Image: ARH modified images from CDC, http://www.cdc.gov/dpdx/malaria/gallery.html#)

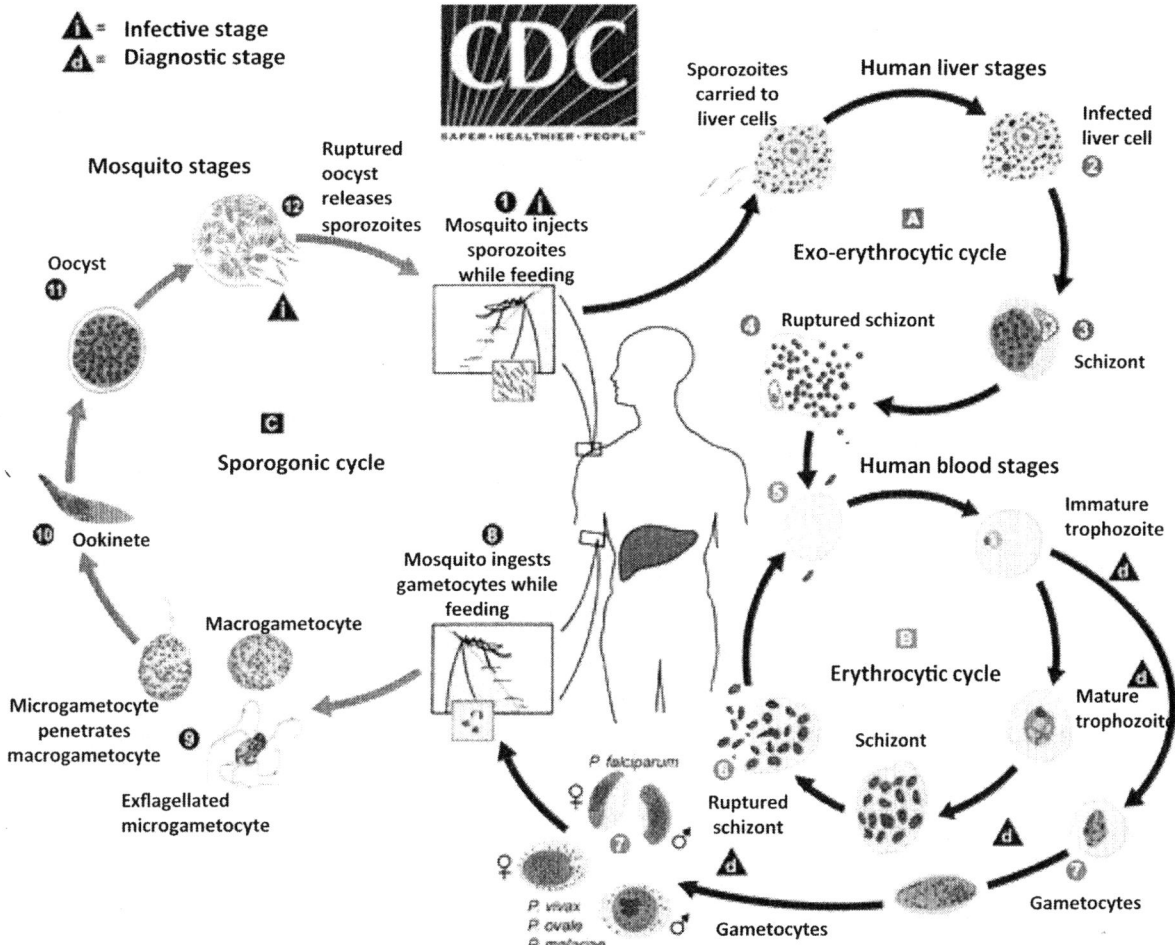

Figure 3.7. Life cycle of *Plasmodium*, the pathogen that causes malaria. The life cycle of *Plasmodium* requires more than one host to complete the life cycle. Infective sporozoites live in salivary glands of *Anopholes* mosquitos. Infection occurs when mosquitos feed and sporozoites are injected into a new human host (1). Within minutes the bloodstream carries sporozoites to liver cells (2). Once inside a liver cell a sporozoite undergoes multiple episodes of binary fission called schizogony and produces thousands of merozoites (also called schizonts) (3). These burst out of liver cells and enter the bloodstream where they penetrate red blood cells (RBCs) and become trophozoites (4-5). Trophozoites also undergo schizogony within RBCs and produce many more merozoites (schizonts) (5-6). These destroy RBCs when they emerge, penetrate other RBCs and become trophozoites. This cycle repeats and produces the symptoms of malaria where a host experiences severe fever and chills every few days. Along the way some trophozoites produce cells called gamonts (gametocytes) (7). Gamonts taken into a mosquito's body with a blood meal produce microgametes and macrogametes that fuse in the mosquito's gut and produce a zygote called an ookinete (8-10). The ookinete burrows through the wall of the mosquito's gut, encysts and produces many sporozoites that migrate to the mosquito's salivary glands when they excyst (11-12). These can be injected into a new host. (Image: ARH modified image from CDC, http://www.cdc.gov/malaria/about/biology/)

Phylum Ciliata/Ciliophora "cilia bearer"

1) Observe any available representatives of this group. Describe and DRAW what you see.
   a. *Balantidium coli* – This is the only known parasitic ciliate that affects humans. It is non-symptomatic in its normal host, swine. Transmission is normally via drinking water or eating food polluted with swine fecal material. Refer to **Fig. 3.8** to help you identify what you see.

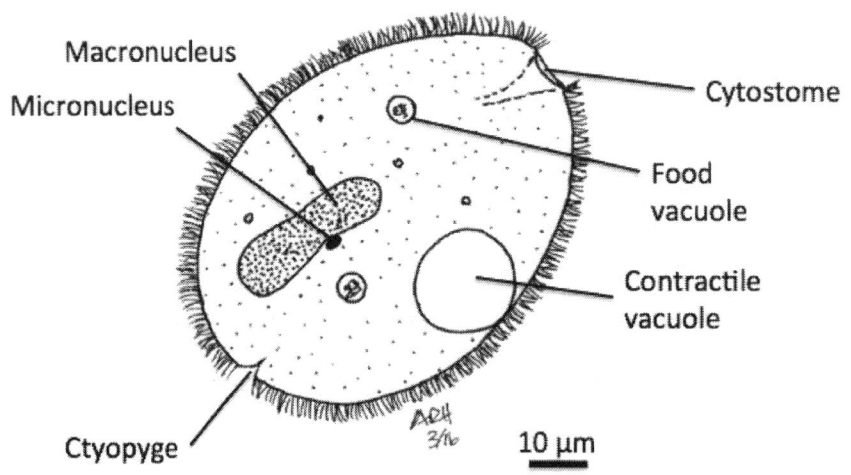

Figure 3.8. *Balantidium coli*, trophozoite stage. This parasitic ciliate causes a condition called balantidiasis. Infection occurs when a host ingests food or water contaminated with fecal material containing cysts. Excystation occurs in the small intestine where trophozoites emerge from cysts and migrate to the lumen of the large intestine. Once in the large intestine trophozoites undergo binary fission and conjugation, some daughter cells move into the lining of the colon and others become mature cysts that are released with the feces. (Image: ARH)

   b. *Didinium sp.* – Free-living heterotrophs found in marine, freshwater and brackish water environments. These prey upon unicellular algae and protozoans. Note the pair of ciliated bands that run around the barrel-shaped body and the pointed oral opening at one end of the body. Refer to **Fig. 3.9** to help you identify what you see. This short video shows *Didinium* feeding: https://www.youtube.com/watch?v=arLutw0b-AY
   c. *Paramecium sp.* – This is a commonly used organism to introduce students to ciliated protozoans. Free-living heterotrophs in marine, freshwater and brackish water. Carries out conjugation (sexual reproduction – see a textbook on invertebrate zoology for a description of this process). Refer to **Fig. 3.10** to help you identify what you see.

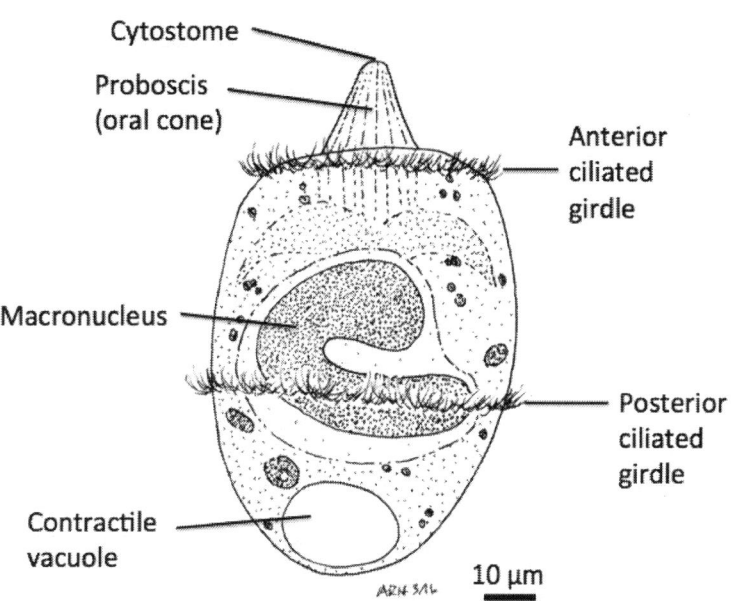

Figure 3.9. The free-living ciliate *Didinium*. Its cytostome expands enough to engulf other protists whole. Engulfing other large cells whole is called holozoic feeding. (Image: ARH)

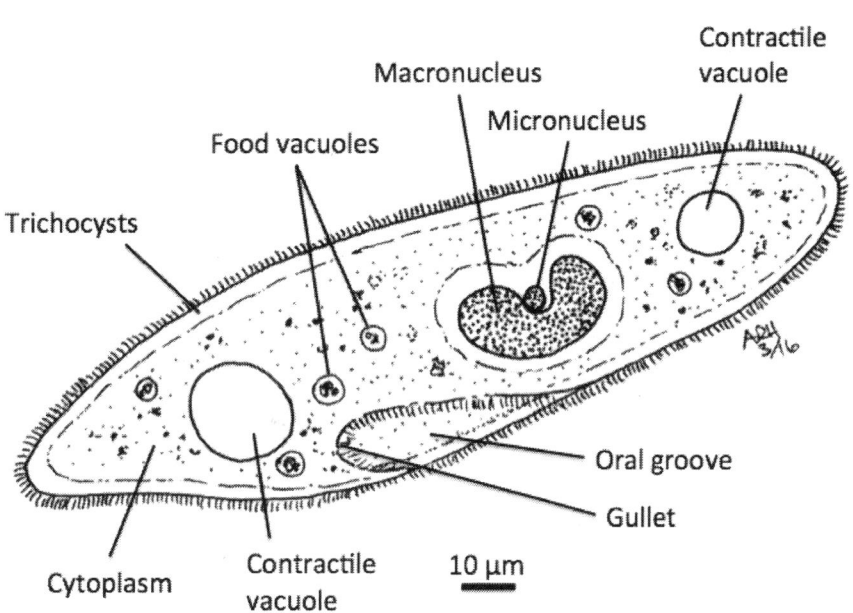

Figure 3.10. The ciliate *Paramecium caudatum*. This protozoan glides through the water using cilia for locomotion. As it does so cilia lining the oral groove pull water, small cells and detritus into the gullet where particles put into food vacuoles. An individual typically has many food vacuoles at any time. This species has a layer of trichocysts just below the outer surface of the plasma membrane. Trichocysts are believed to be mainly defensive in purpose and are harpoon

or nail-shaped organelles that can be fired when *Paramecium* is threatened or experiences the right kind of stimulus. (Image: ARH)

    d. *Spirostomum sp.* – Free-living, large (to 4mm in length), elongate heterotrophic species with longitudinal rows of cilia. Found in freshwater and marine habitats. Refer to **Fig. 3.11** to help you identify what you see.

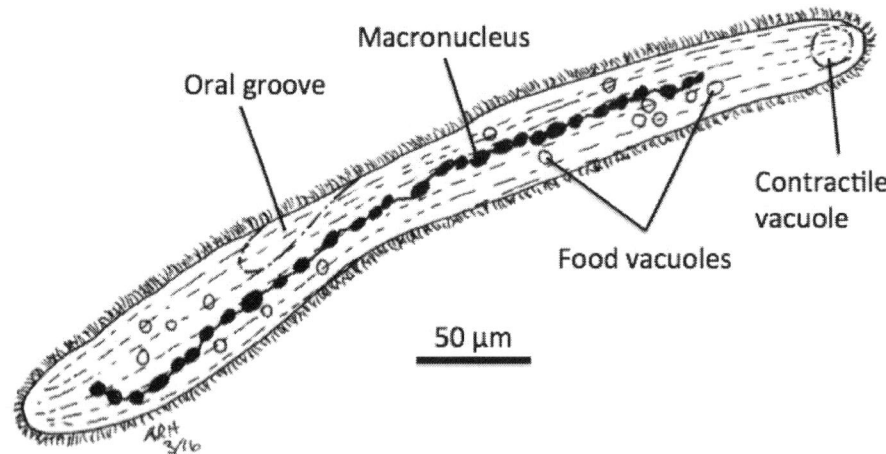

Figure 3.11. The large ciliate *Spirostomum sp.* This elongate ciliate has many rows of cilia along its length. It feeds as cilia lining the oral groove pull in water, small cells and detritus. Food particles are put into food vacuoles where they can be digested. There is a large contractile vacuole at the posterior end of the cell that can take up a large portion of the posterior part of the cell. The macronucleus sometimes looks like beads on a string. (Image: ARH)

Figure 3.12. The ciliate *Stentor sp.* This is an active protist. It uses contractile myonemes in its membrane to contract its body when it is perturbed. The band of cilia at the anterior end of the body creates a feeding current that brings food particles toward a cytostome where they can be ingested and put into food vacuoles. There is typically a large contractile vacuole near the anterior end of the body, and the macronucleus can look like beads on a string. These organisms can detach themselves and swim freely

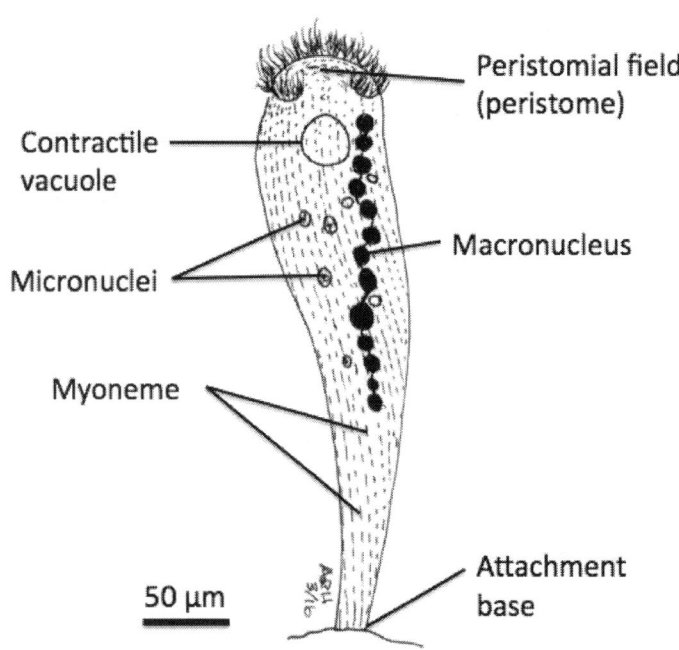

as needed and then reattach themselves. (Image: ARH)

> e. *Stentor sp.* – Trumpet-shaped suspension feeders that use their cilia to obtain food by generating currents of water that bring suspended particles in contact with the oral groove of the body. Found primarily in freshwater habitats, rarely in marine systems. A large protozoan that can reach 2mm in length. Refer to **Fig. 3.12** to help you identify what you see.
>
> f. *Vorticella sp.* – Sessile cone-shaped ciliates with a thin stalk containing a coiled contractile myoneme. When the myoneme contracts it causes the stalk to coil tightly and quickly pulls the main body of the cell downward toward the substrate. Like *Stentor*, *Vorticella* uses its cilia to generate water currents that bring suspended particles close enough to be captured. Refer to **Fig. 3.13** to help you identify what you see.

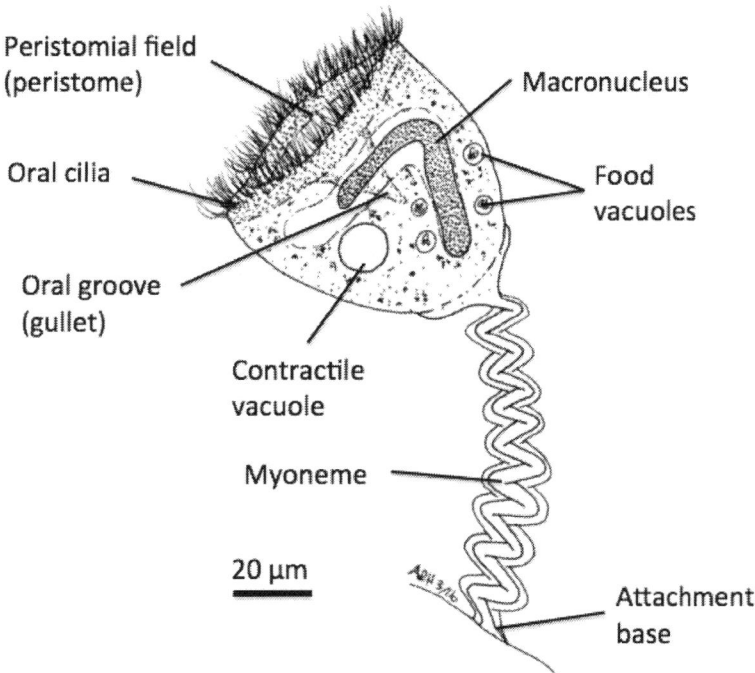

Figure 3.13. The ciliate *Vorticella sp.* This protist is sessile. The stalk that attaches the main body of the cell to the substrate contains a contractile myoneme. When it contracts it coils up like an old-style phone cord and it can also be extended fully so there is no coiling visible. The contraction of the myoneme is extremely fast. This organism uses oral cilia surrounding the peristomial field to generate a feeding current that pulls particles of food like bacteria into the oral groove (gullet) where they are engulfed and placed in food vacuoles. (Image: ARH)

Task #3 – Group 3 Rhizaria

Phylum Granuloreticulosa "little grain network"

1) Foraminferans - Commonly called "forams", these protozoans secrete ornate calcium carbonate tests often reminiscent of tiny, coiled snail shells but bearing many small openings through which their long thin pseudopods (filopodia, reticulopodia) are extended. These are free-living heterotrophs that use their pseudopodia to phagocytize organic material. Observe, describe and DRAW any specimens available for study. The

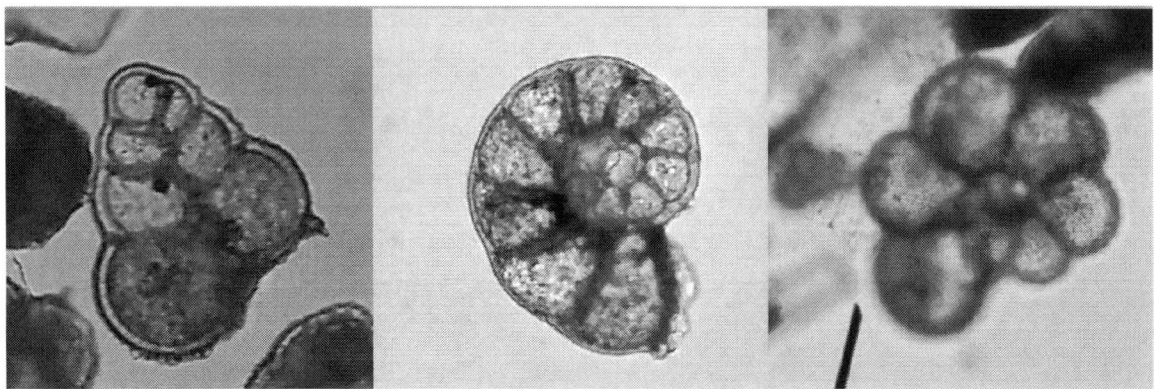

sizes and shapes of foraminiferan tests (shells) can be highly variable. A few forms are included in **Fig. 3.14**.

Figure 3.14. A few examples of foraminiferan tests. (Images: ARH)

Phylum Radiolaria "spoke-like"

2) Radiolarians – These single-celled protozoans secrete ornate silicon dioxide tests bearing many small openings. Radiolarians extend long, thin pseudopods (filopodia, reticulopodia) through these openings. These are also free-living phagocytic heterotrophs. Observe, describe and DRAW what you see. A few examples of radiolarian tests are shown in **Fig. 3.15**.

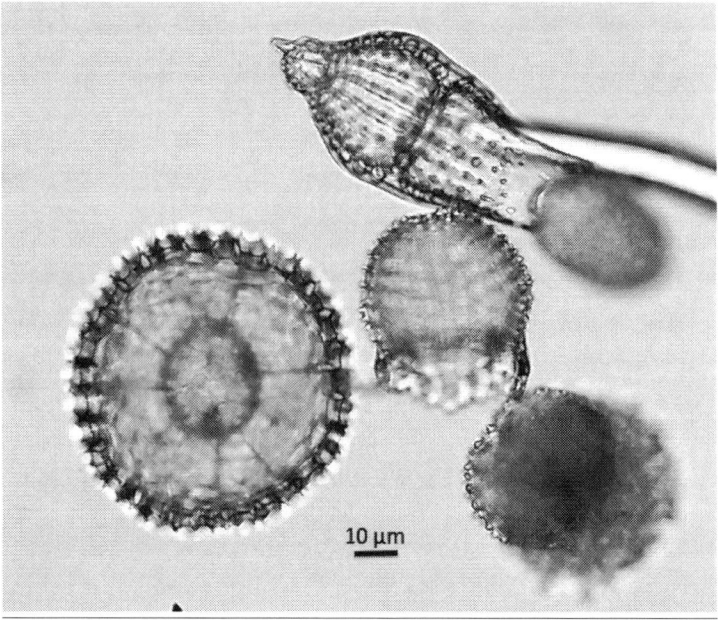

Figure 3.15. Radiolarians. (Image: ARH)

Task #4 – Group 4 Excavata "out of a hollow or depression"

Phylum Parabasalida "beside the base"

1) Observe the trichomonad *Pentatrichomonas (Trichomonas) hominis* and DRAW what you see (**Fig. 3.16**). This protozoan is not known to cause disease in humans and little is known about its biology. According to the CDC, hosts become infected when they ingest food or water that is contaminated with feces or fomites (other objects capable of carrying pathogens). Trophozoites live in the large intestine but are considered commensals in humans.

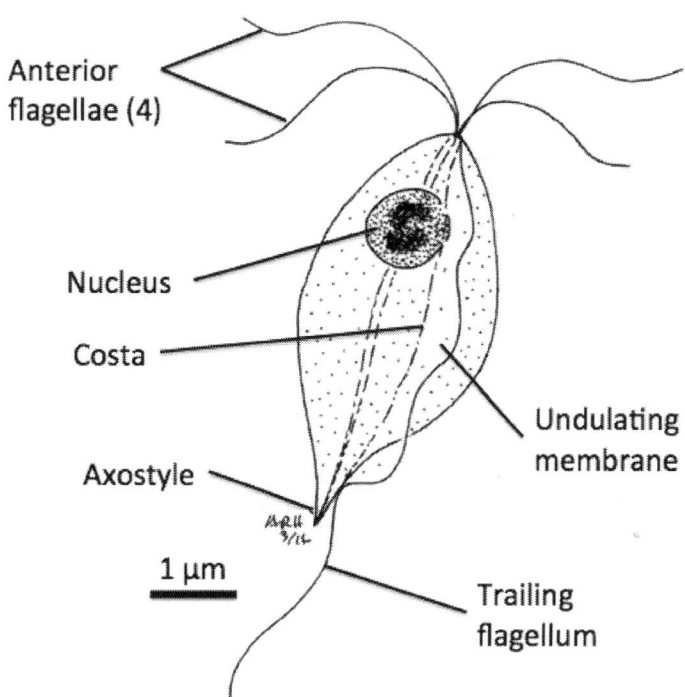

Figure 3.16. Trophozoite of *Pentatrichomonas (Trichomonas) hominis*. There are typically four anterior flagellae and one trailing flagellum that is connected to the cell along most of its length via an undulating membrane. These protozoans are quite small and it is difficult to see many anatomical structures of this species even when observed using the 100x oil immersion objective lens. (Image: ARH)

Phylum Diplomonadida "two singles"

2) Observe *Giardia lamblia* (*intestinalis*) and DRAW what you see, **Fig. 3.17**. This species causes giardiasis, a condition commonly referred to as backpacker disease. Non-human hosts can include deer, sheep, beavers, cattle, etc. Cysts are released in the feces of an infected individual and can survive for prolonged periods of time in cold water. Humans are infected when they drink water or eat food contaminated with cysts. Excystation occurs in the small intestine and a trophozoite is released. These remain in the small intestine and undergo binary fission. Trophozoites can either remain free in the lumen of the intestine or use their adhesive discs to attach themselves to the mucosa of the gut. Cysts are formed as trophozoites move toward the colon. According to the CDC, cysts are infective shortly after release so person-to-person transmission is possible.

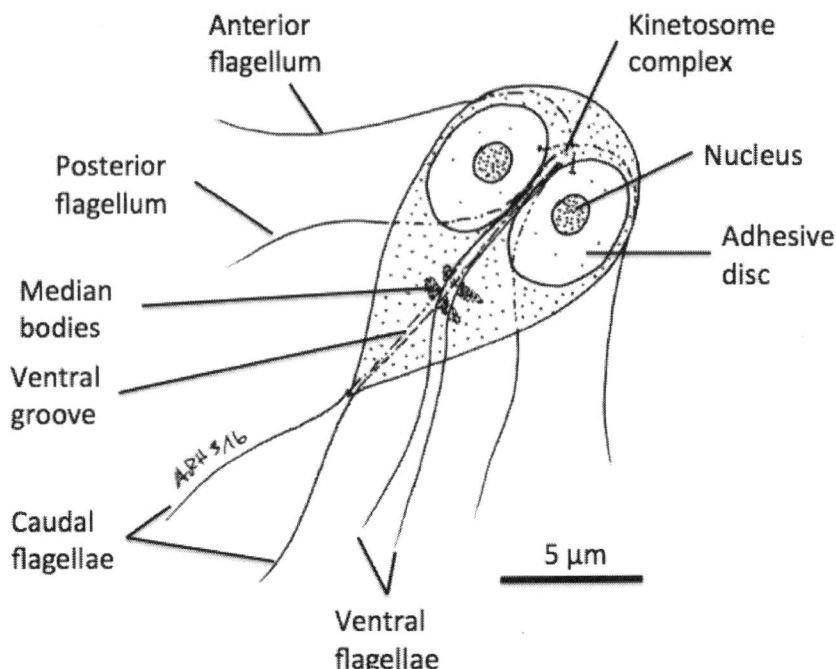

Figure 3.17. Trophozoite of *Giardia lamblia* (*intestinalis*). The anterior, posterior and caudal flagellae have segments that are internal and segments that are external to the cell. All flagellae originate in the kinetosome complex. Median bodies apparently contribute to the proper formation of the adhesive discs, without median bodies the discs do not form correctly. You will need to use the 100x oil immersion lens on your compound microscope to see these organisms. (Image: ARH)

Phylum Euglenida "true pupil (eyeball)"

3) Euglenozoa – *Euglena*, **Fig. 3.18.** Members of this group are mixotrophs, organisms that carry out photosynthesis under some conditions and heterotrophy under other conditions. They have multiple chloroplasts and store excess nutrients they produce via photosynthesis as a starch-like compound called paramylon. They also have a heavily pigmented structure called an eyespot or stigmata. This structure does not sense light; instead the cell reorients itself so the eyespot shields the reservoir and cellular apparatus that run the flagellae when light intensity is too great. *Euglena* has two flagellae: one of them extends beyond the plasma membrane and is used for locomotion and the other one is quite short and is housed completely inside the reservoir. *Euglena* is an important indicator organism of eutrophic (polluted) water.

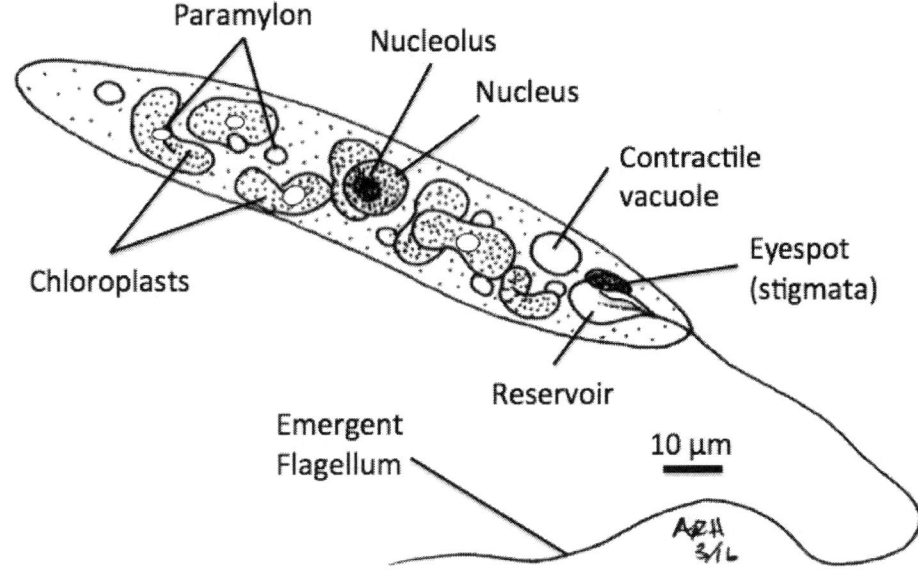

Figure 3.18. *Euglena*, a mixotrophic organism. (Image: ARH)

Phylum Kinetoplastida "moving cell"

4) *Trypanosoma* – members of this genus cause the fatal diseases Chagas disease and African sleeping sickness. These tiny flagellates live in the blood. **Figure 3.19** shows individuals in a blood smear preparation. **Fig 3.20** shows the life cycle of these parasites. You will need to use the 100x oil immersion lens to see these parasites.

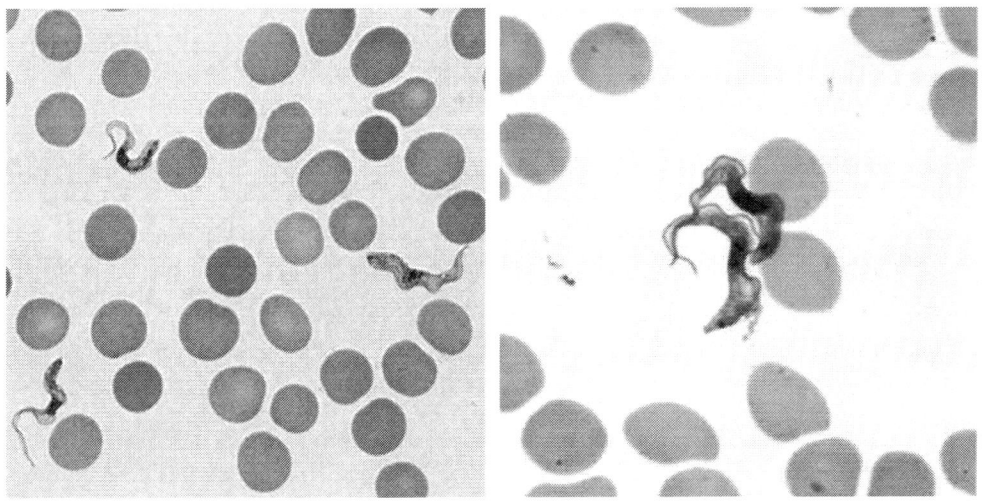

Figure 3.19. The flagellate *Trypanosoma* in a blood smear preparation; oblong objects are RBCs. (Image: ARH modified images from CDC, http://www.cdc.gov/dpdx/trypanosomiasisAfrican/gallery.html)

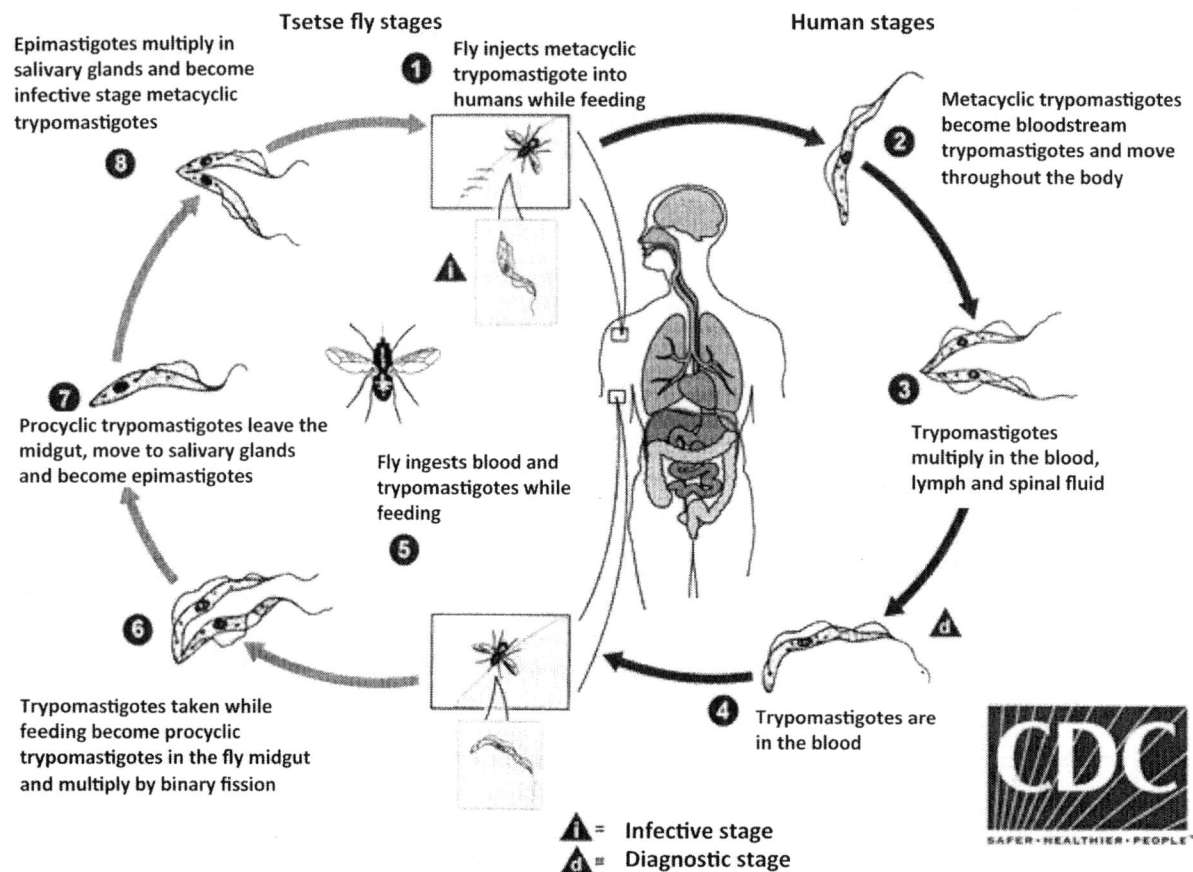

Figure 3.20. The life cycle of *Trypanosoma brucei*, a species that causes African sleeping sickness. The life cycle begins when an infected tsetse fly bites a human and injects saliva containing metacyclic trypomastigote-stage (infective stage) individuals into the skin (1). These move from the lymphatic system to the blood stream where they transform into bloodstream trypomastigotes (2-4). These are carried throughout the body and invade other body fluids including lymph and spinal fluid where they continue replication by binary fission. All life stages are extracellular. A tsetse fly becomes infected when it feeds on the blood of an infected individual (5). Once in the fly's midgut, parasites become procyclic trypomastigotes that multiply by binary fission (6). These leave the midgut and become epimastigotes (7). Epimastigotes invade the fly's salivary glands where they continue multiplication by binary fission and await injection into a new host (8). It takes about three weeks for parasites in a fly to be ready to infect a new host. Symptoms of African sleeping sickness occur when these parasites cross the blood-brain barrier and produce neurological changes in the CNS that are evidenced by general lethargy and sleepiness and premature death. (Image: CDC, http://www.cdc.gov/parasites/sleepingsickness/biology.html)

Task #5 – Group Opisthokonta "posterior flagellum"

Phylum Choanoflagellata "funnel flagellum"

1) Observe any available choanoflagellate (**Fig. 3.21**). Choanoflagellates are currently identified as the most likely sister taxon to clade Metazoa. Choanoflagellates look quite similar to choanocytes of sponges. They both have a collar of microvilli surrounding a single flagellum. DRAW what you see.

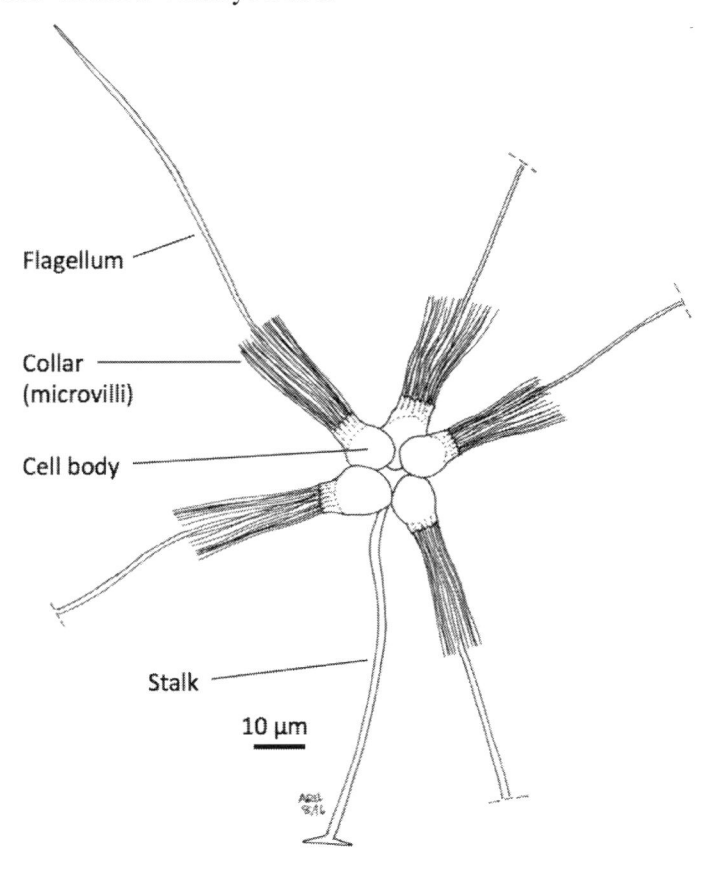

Figure 3.21. A colonial choanoflagellate. (Image: ARH, after an SEM image in http://www.dayel.com/choanoflagellates/, creative commons share-alike access though https://creativecommons.org/licenses/by-sa/3.0/)

Group questions:

1) What does it mean when we say that a taxon is polyphyletic?
2) What particular osmotic challenge do freshwater unicellular organisms face?
3) Why do we currently identify choanoflagellates as the sister taxon of Kingdom Animalia?

Eukarya – Glossary

Alveoli – tiny sac-like structures associated with the plasma membrane in some protists

Autotroph – an organism that can produce its own food either via photosynthesis or chemoautotrophy

Axostyle – in Trichomonads, short pointed structure that extends beyond the point where the flagellum of the undulating membrane becomes the trailing flagellum, function unknown

Coccidiosis - disease affecting the intestines of birds and mammals caused by some protozoan parasites, e.g., *Eimeria*

Contractile vacuole – organelle used to collect and expel excess water

Costa – in Trichomonads, where the undulating membrane attaches to the plasma membrane

Cytopyge – location where waste material is released from some protists

Cytostome – location where food, etc., is taken into some protists

Ectoplasm – more viscous (thicker) clear outermost portion of the cytoplasm

Endoplasm – fluid portion of cytoplasm that contains granules and cytoplasmic organelles

Epitheca – in dinoflagellates, the portion of the protective outer covering (theca) above the equatorial girdle

Excystation – when an infective stage protist parasite emerges from its protective cyst

Eyespot (stigmata) – pigmented structure that can shade the reservoir in *Euglena* as needed to protect it from excessive light

Filopodia – thread-like pseudopodia used by some protists to capture and ingest food

Flagellum (ae = plural) – whip-like organelle used for locomotion or food collection

Food vacuole – membrane-bound vesicle containing food until it is digested

Gamont/gametocyte – parasitic life stage capable of producing gametes

Girdle – in dinoflagellates, the groove in the theca that runs around the middle of the cell

Gullet – internal tube-like structure attached to the oral groove on one end and the cytostome on the other end

Heterotroph – an organism that cannot make its own food and uses organic material produced by other living things for nutrition

Hypotheca – in dinoflagellates, portion of the theca below the girdle

Kinetosome – basal body supporting a flagellum

Lobopodia – broad, rounded pseudopodia used for locomotion and feeding

Macrogamete – larger gamete, usually considered to be the female gamete

Macronucleus – polyploid nucleus that controls most of the non-reproductive functions of a cell

Median bodies – dark inclusions in the cytoplasm of *Giardia*, contribute to formation of the adhesive discs

Merozoite – parasitic life stage that can clone or carry out sexual reproduction

Microgamete – smaller gamete, usually considered to be the male gamete

Micronucleus – nucleus that contains genetic material involved in sexual reproduction

Mixotroph – an organism that can either make its own food or use organic material made by other organisms for nutrition

Myoneme – longitudinal contractile structure composed of protein filaments

Nucleolus – dense spherical body found in the nucleus during interphase, its main function is to produce subunits of ribosomes

Oral groove – external groove or depression in some protists that lead to an internal gullet and cytostome

Paramylon – starch-like carbohydrate

Parasite – organism that feeds on a host organism without killing it (at least right away), can either live outside (ectoparasite) or inside (endoparasite) the host

Peristomial field – area of the plasma membrane surrounded by a band of cilia and often bearing a cytostome

Plastid – term used to refer to chloroplasts and a few other kinds of membrane bound organelles

Proboscis – extension of the plasma membrane in some protists that bears the cytostome

Pseudopodia – extensions of the plasma membrane used for movement and food capture

RBC – red blood cell

Reservoir – in *Euglena*, an enclosed space where the two flagellae are inserted into the plasma membrane

Sulcus – in dinoflagellates, the groove of the hypotheca that runs down from the girdle and houses the longitudinal flagellum

Test – hard outer covering

Trichocyst – tiny inclusion in the plasma membrane of some protists that can fire like small lances or harpoons

Trophozoite – feeding stage of some parasitic protozoans

Vacuole – membrane-bound organelle may contain fluid or fluid and food, etc.

Chapter 4: Phylum Porifera

All animals belong to Kingdom Animalia/Clade Metazoa. Metazoans are strictly multicellular, heterotrophic, lack cell walls, motile at some stage of the life cycle, produce embryonic tissue layers and their development is directed at least in part by developmental regulatory genes called *Hox* genes (short for homeobox genes). The sister taxon of Clade Metazoa is the protist phylum Choanoflagellata.

Porifera "pore bearers" includes about 9,000 species; all but about 200 species live in marine habitats. Sponges have been around for more than 500 million years. They have anatomically simple bodies and are recognized as the basal animal taxon, i.e., the taxon at the base of the animal phylogenetic tree. Sponges lack many animal traits such as integrated neuromuscular systems, multi-tissue organs, cell gap-junctions, body polarity and division of labor at the level of tissues and organs. Sponges also lack some characteristic *Hox* genes. They do however carry out sexual reproduction and cleavage events produce embryos with layered tissues. Characteristics of sponges are listed in **Table 4.1.**

Table 4.1 Characteristics of Phylum Porifera (after Brusca, *et al.*, 2016).

- Cellular grade construction, some species have adhaerens producing simple tissues but lack gap junctions
- Embryos produce tissue layers and express body polarity but adults are asymmetrical and lack any evidence of body polarity
- Cells retain multipotency or pleuripotency
- Flagellated choanocytes (collar cells) used for generating water currents and feeding
- Type IV collagen (major structural component of basement membranes in animals) common in basal membranes of some sponges
- The mesohyle layer between inner and outer layers of the body wall always contains motile amoeboid cells and usually contains support structures such as spicules or spongin
- Highly unusual in producing both cilia (monociliated cells – embryonic epidermal cells and cells lining oscula) and flagellae (choanocytes)

Sponges are anatomically simple but are ecologically complex. Most animals deal with ecological challenges via complex behavioral responses but sponges can't do this. All they can do is sit there attached to a rock or something while they deal with challenges like predation or competition. Though sponges don't employ complex behaviors they are masters of chemical warfare. Biologically active molecules produced by sponges have been discovered that are anti-viral, anti-bacterial, anti-fungal, immunosuppressant and neuroinhibitory among others. These chemicals allow sponges to fend off most pathogens, predators and competitors.

Sponge taxonomy has undergone significant changes recently and results show that sponges are monophyletic. All sponges are members of Phylum Porifera that contains four classes. Class Calcarea has fewer than 700 species and is comprised of sponges that produce only calcium carbonate spicules. Class Hexactinellida are the glass sponges, including about 700 species, all of which produce tri-axon or six-rayed silicon dioxide spicules and have syncytial soft-tissue construction. Class Demospongia is the largest class of sponges with over 7,000 species. All of these have cellular construction and produce silicon dioxide spicules or an organic skeleton of the protein spongin. Class Homoscleromorpha is a recently proposed taxon of fewer than 100 species. These sponges lack spongin but always produce tetra-axon spicules when spicules are present and all spicules are the same size.

Sponge taxonomy is notoriously difficult so this exercise does not present a class-by-class approach to the sponges, though some classes are indicated. Instead this exercise gives you an opportunity look at some of the different body plans found within this phylum. Sponges generally exhibit three different body plans: asconoid, scyonoid and leuconoid. These body plans are not exclusive to any class and are therefore not particularly useful when it comes to resolving taxonomic questions. Understanding differences between these body plans however, provides useful insights into how sponges work.

Watch this 14-minute video from the series *The Shape of Life*. It provides a good overview of the body plan and origins of sponges: http://www.shapeoflife.org/video/sponges-origins.

Class Hexactinellida "small six-rayed" – glass sponges

Members of this class are relatively easy to identify because of their extensive and elaborate internal skeleton. *Euplectella* is a representative genus in this class. It lives in a highly unusual habitat for a sponge, the deep-sea muddy environment. Most sponges require a hard substrate for attachment and survival but this sponge produces a highly complex endoskeleton comprised of interconnected SiO_2 spicules that allows it to survive on soft substrates. One end of the tubular body bears the excurrent opening (osculum) and the opposite end produces many long hair-like spicules used to anchor the sponge in soft sediments.

If you have a chance to observe a skeleton of one of these sponges first-hand, check to see if there are remains of *Spongicola spp.* shrimp in the spongocoel (see **Fig 4.3**). These shrimp, usually one male and one female, take up residence inside the spongocoel of a hexactinellid sponge when they are small enough fit through small openings in the osculum. As the shrimp grow they become too large to escape and they live the rest of their lives inside the sponge. Their offspring escape the sponge when they are still small enough to fit through the openings in the cap of the osculum. Offspring then search for their own unoccupied glass sponges where they can take up residence.

Glass sponges differ from most other sponges in that they are syncytial, i.e., their cells are multinucleate and only a few cells or even one cell can cover the entire skeleton.

Task #1 – glass sponges:

1) Examine and DRAW an entire skeleton of *Euplectella*. Refer to **Fig. 4.1** to help you identify what you see.

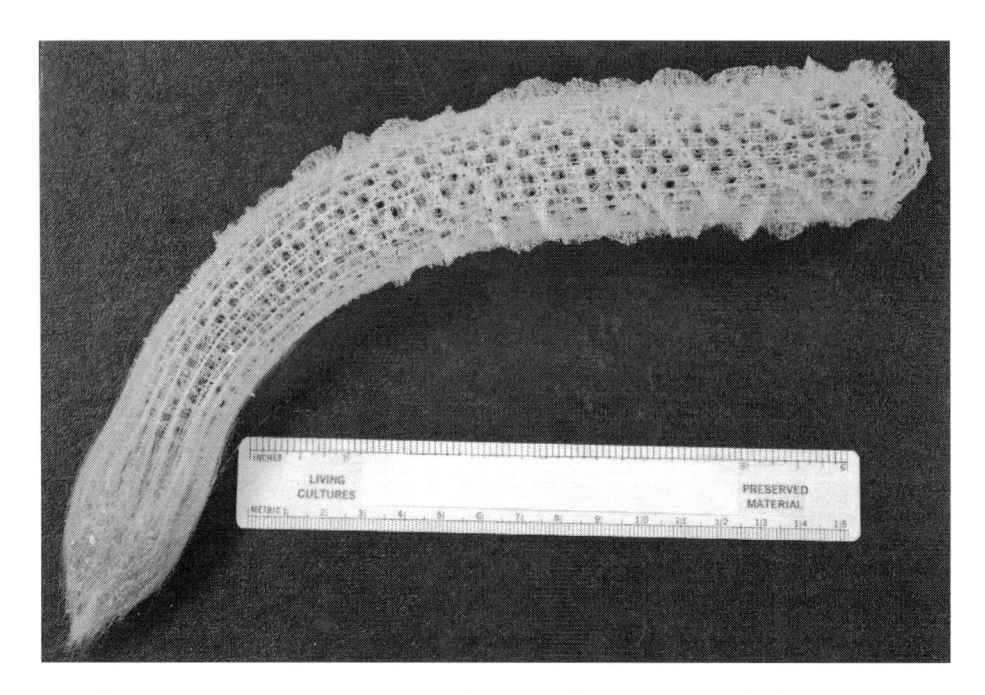

Figure 4.1. Skeleton of the glass sponge *Euplectella*. (Image: ARH)

2) Use a magnifying lens or dissection scope to study the skeleton of *Euplectella* in greater detail. If you do not have access to a skeleton of *Euplectella* search for photographs of this interesting species online and use **Fig. 4.2** to complete this portion of the lab. DRAW a small section of the spicule body wall. Don't forget to include a scale bar and observations and questions to accompany every entry in your lab notebook.

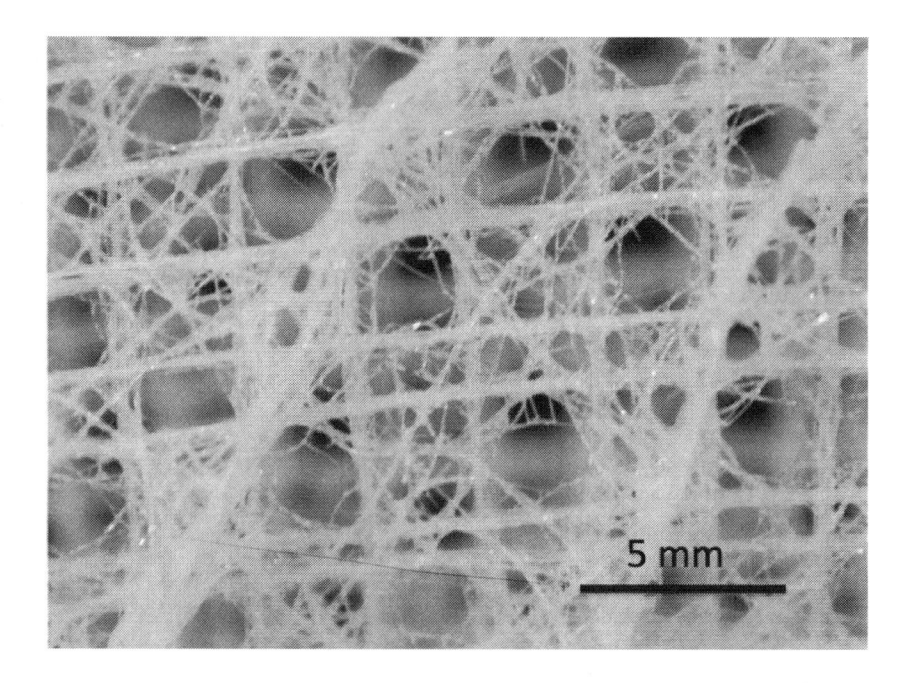

Figure 4.2. Spicules in the wall of the glass sponge *Euplectella*. (Image: ARH)

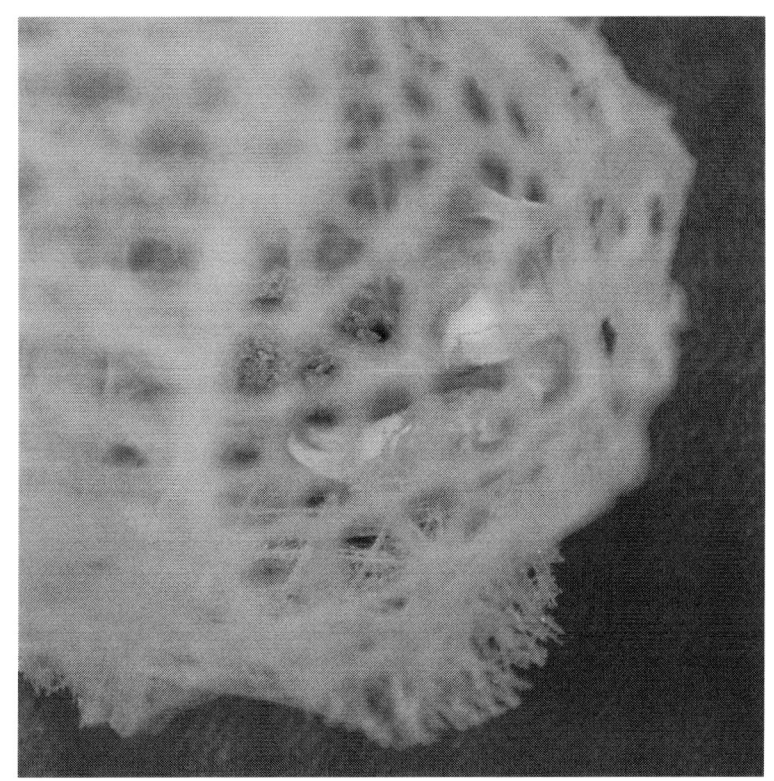

Figure 4.3. Remains of the shrimp *Spongicola* visible in the spongocoel of a skeleton of *Euplectella*. (Image: ARH)

Asconoid body plan "sac-like"

Leucosolenia, a genus in Class Calcarea includes sponges that have the asconoid body plan. The asconoid plan is considered to be the least anatomically complex of all sponge body plans. It may not look particularly simple when you see a mass of interconnected tubes arising from a common base but each finger-like projection of the colony has a thin body wall separating the external environment from the choanocyte-lined inner chamber known as the spongocoel or atrium. Water enters the body through many tiny openings called ostia that are scattered across the body wall. Ostia are too small to be seen even with a compound microscope. Water is pushed and pulled through the spongocoel and eventually leaves the sponge body via one of many visible excurrent openings called oscula that are located at the tip of each finger-like extension of the body. As the water passes through the choanocyte-lined spongocoel choanocytes phagocytize even the tiniest particles of organic matter suspended in the water.

The collective beating of choanocytes and passive water flow produced by the Bernoulli effect resulting from differing water flow rates at the basal boundary layer of the colony and faster flowing water passing the oscula at the tip of each branch allows the sponge to process impressive volumes of water. A vertical distance of even a few millimeters can be enough to produce passive water flow.

Task #2 – asconoid body plan:

1) Immerse a small part of a cluster of preserved or live *Leucosolenia*. Observe the sponge's shape with a magnifying glass or dissection microscope. DRAW a small part of the cluster and be sure to include at least one osculum in your drawing (see **Fig. 4.4**).

Figure 4.4. A small cluster of the sponge *Leucosolenia.* (Image: ARH)

2) Make a wet-mount of one branch of *Leucosolenia*. Observe prepared slides if live or preserved specimens are not available. DRAW some spicules visible in the body wall (see **Fig. 4.5**).

Figure 4.5. Triaxon spicules in the body wall of *Leucosolenia.* (Image: ARH)

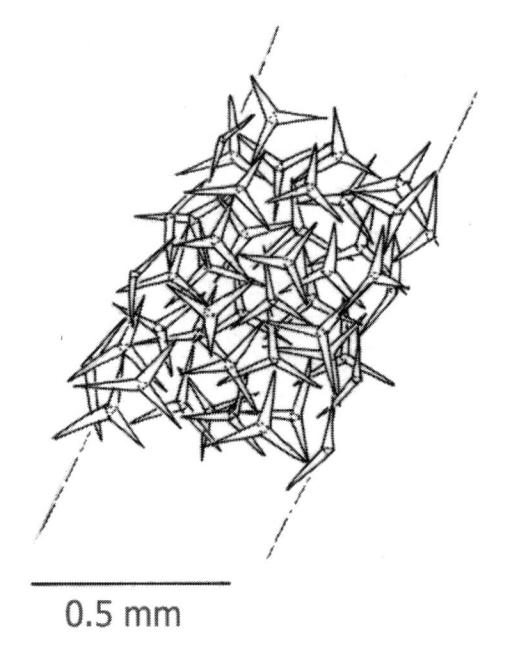

3) Take a small sample of spicules and add a drop or two of acid to them. Describe how the acid affects the spicules. How does what you observe mesh with the fact that this sponge belongs to Class Calcarea? WRITE your observations in your laboratory notebook.

Syconoid body plan "fig like"

Scypha and *Grantia,* also members of Class Calcarea are two commonly available genera of sponges that exhibit the syconoid body plan. This body plan is more anatomically complex than the asconoid plan. The flow of water through this body plan occurs as follows: 1) water passes ostia and enters incurrent canals; 2) water passes from the incurrent canal through another opening called a prosopyle which empties into radial canals that are lined by choanocytes, i.e., choanocyte chambers; 3) water then passes through apopyles, the excurrent openings at the downstream end of radial canals and enters the spongocoel and; 4) water leaves the sponge via an osculum.

The radial canals in the finger-like extensions from the body wall allows sponges with the scyonoid plan produce a much larger cross sectional area than sponges with the asconoid plan.

Task #3 – syconoid body plan:

1) Immerse a specimen of *Scypha* or *Grantia* and observe it with a magnifying glass or dissection scope. DRAW the whole sponge (see **Fig. 4.6**).

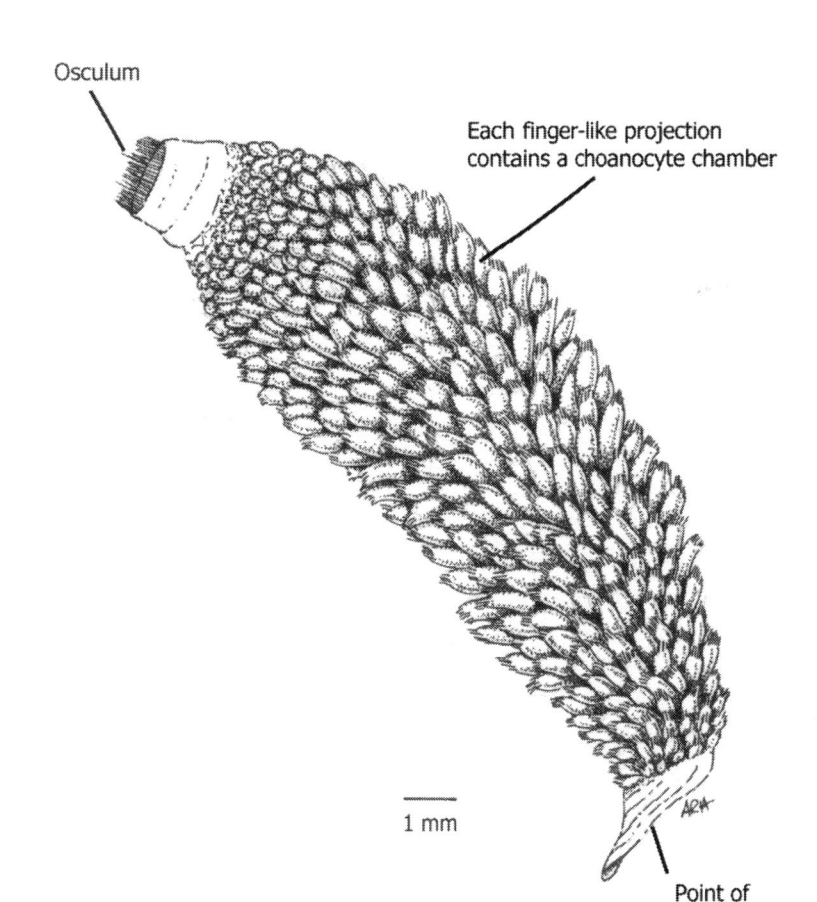

Figure 4.6. *Grantia,* whole body. (Image: ARH)

2) Observe a cross-section slide of *Scypha* or *Grantia* that contains eggs and another that contains embryos. DRAW a composite sketch of these slides showing one radial canal bearing eggs and another one bearing embryos. Refer to **Fig.4.7** for a model of what this might look like and to help you identify what you see.

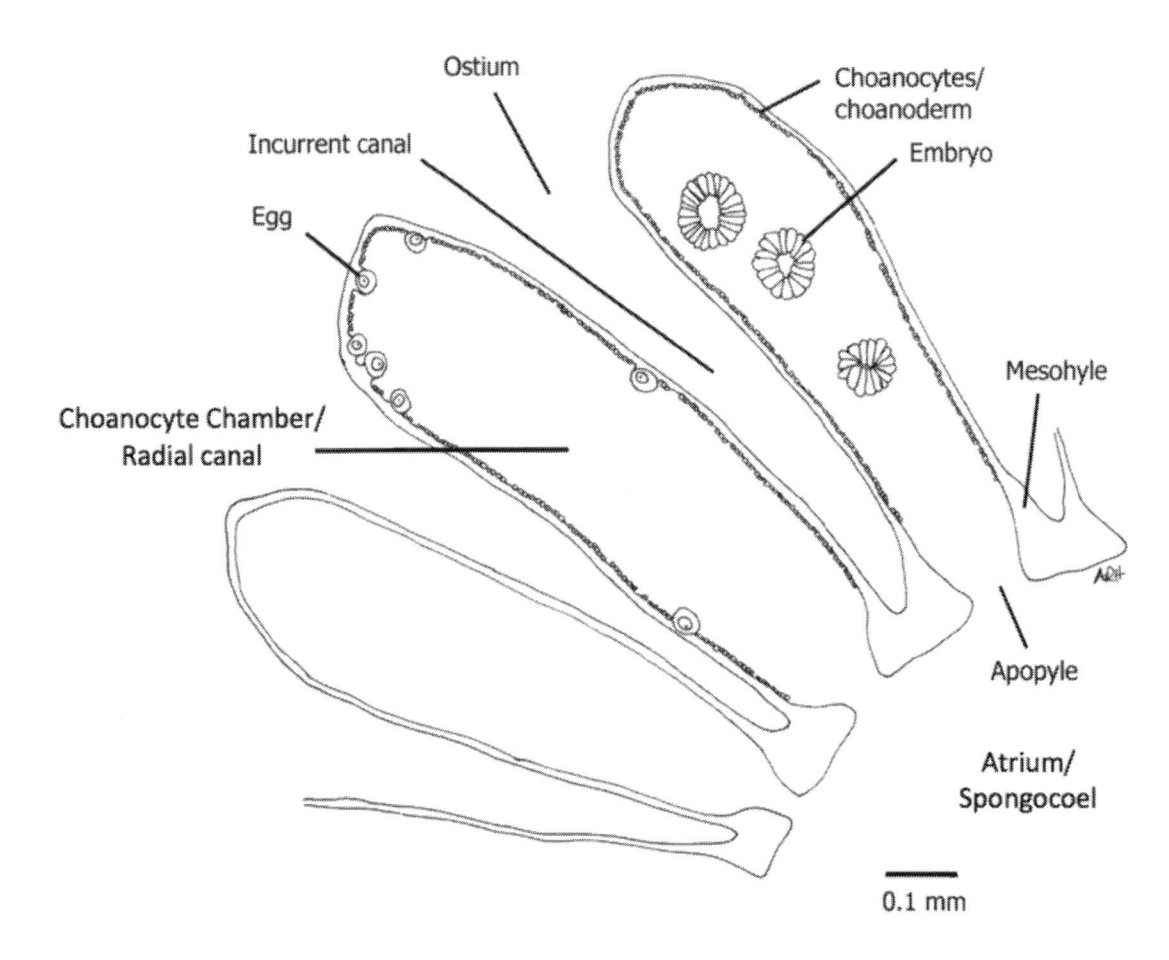

Figure 4.7. Composite drawing of cross-section views though a syconoid the sponge showing one choanocyte chamber bearing eggs and another one bearing embryos. (Image: ARH)

Leuconoid body plan "throat like"

The vast majority of sponges produce the leuconoid body plan, the most anatomically complex body plan of sponges. Leuconoid sponges typically have spongin as part or all of their skeletal support. Spongin is a collagen-based protein that is stiff yet flexible and gives bath sponges their spongy characteristics.

You may already be familiar with these sponges. Sponges with this body plan include natural bath sponges some people like to use to wash their cars as well as elephant ear sponges used in pottery making. Most of the sponges you will spot during low tide also have this plan.

In the leuconoid plan water enters the sponge through ostia that lead to a series of branching incurrent canals. These canals become progressively smaller in cross-sectional area and eventually empty into tiny spherical choanocyte-lined feeding chambers. How tiny? There can be as many as 18,000 choanocyte chambers per mm^3 in leuconoid sponges. Water exits these choanocyte chambers via excurrent canals that fuse with each other to form progressively larger canals that eventually empty into a spongocoel and water exits the body through one of many oscula.

The collective cross-sectional area of different parts of the system of water canals determines local flow rates in these animals. For example the collective cross-sectional area of all ostia is much smaller than the collective cross-sectional area of all choanocyte chambers. This means that according to the Bernoulli Equation that water flow is faster as the same volume of water flows through ostia than when the same volume passes through choanocyte chambers. It's like the same volume of water passing through a small diameter pipe versus a larger diameter pipe; it moves faster through the smaller than the larger pipe. Because the cumulative cross-sectional area choanocyte chambers in a sponge is so much larger than the cumulative cross sectional area of any other part of the water canal system this is where water flows the slowest. This is a great strategy for sponges since these chambers are where they capture the tiniest particles suspended in the water. Water flow rates increase again as water leaves choanocyte chambers, moves into fewer and fewer larger canals until water is expelled through oscula.

Task #4 – leuconoid body plan:

1) Observe dried sponges with the naked eye and a magnifying lens or a dissection microscope. WRITE your observations in your laboratory notebook.
2) Make a wet mount slide of the smallest piece of spongin you can remove from a bath sponge. Observe it under a compound scope. DRAW some of the spongin (see **Fig. 4.8**) and WRITE about how the structure of spongin gives bath sponges their spongy characteristics.

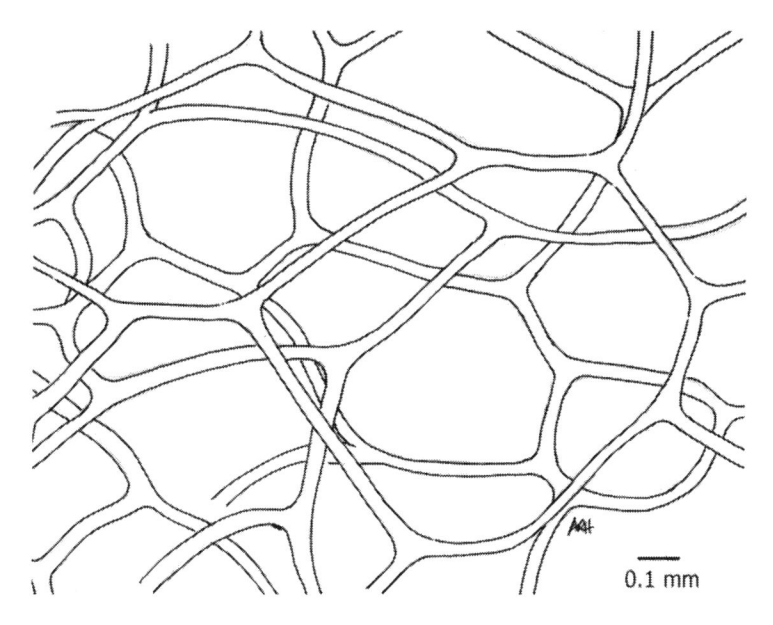

Figure 4.8. Spongin from a bath sponge. (Image: ARH)

3) Observe preserved specimens as well as mounted slides of the freshwater sponge *Spongilla*. DESCRIBE what about what you see.
4) Observe a prepared slide of gemmules from a freshwater sponge. Gemmules are

overwintering bodies – masses of cells called archaeocytes/thesocytes that are protected by a layer of densely packed spicules. Gemmules are not embryos or reproductive cells. They contain a mass of undifferentiated cells that emerge when conditions improve and form a new sponge body. Freshwater sponges produce gemmules when environmental conditions deteriorate, usually at the onset of winter. DRAW a gemmule and use **Fig. 4.9** to help you identify what you see.

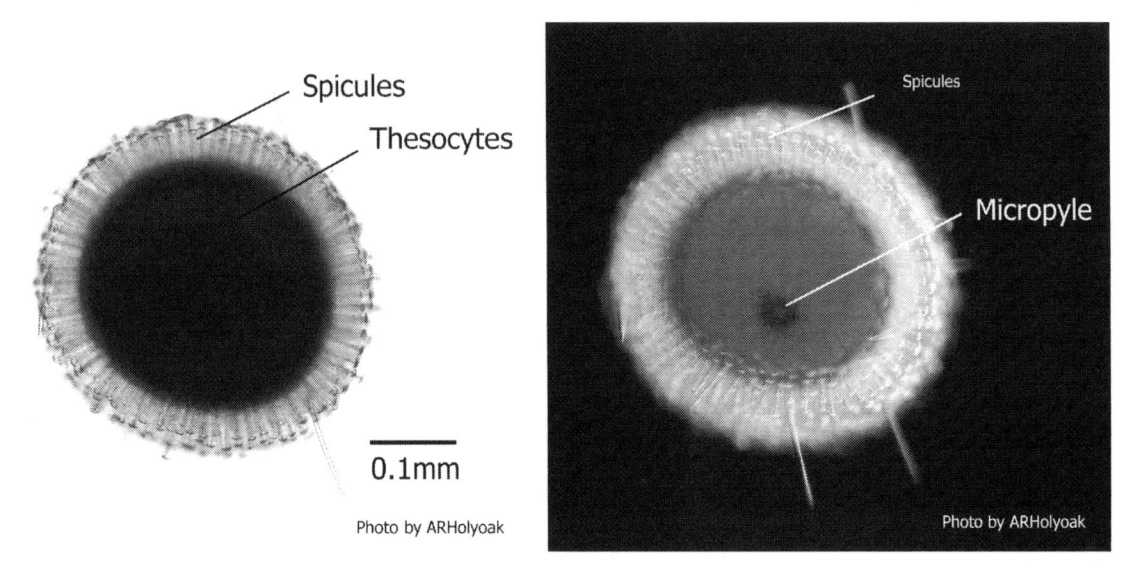

Figure 4.9. Gemmule from a freshwater sponge: bright field image (left), and dark field (right). (Image: ARH)

Group Questions

The current paradigm states that sponges were the first animals. There were no predators when sponges evolved and available food was floating in the water. Keep these things in mind as you answer the following questions:

1) Consider origins of sponge skeletons, i.e., what environmental pressures could have led to the production of sponge skeletons? Develop one hypothesis that includes predation pressure as a driving force and one hypothesis that does not.
2) Based on what you now know do you agree or disagree with treating sponges as only marginal animals by referring to them as parazoans (alongside animals) or do you think they should be considered to be full-fledged animals?

Phylum Porifera – Glossary

Adherens – cell-to-cell adhesive proteins produced by animals including tight junctions, gap junctions and desmosomes

Apopyle – excurrent opening leaving a choanocyte chamber

Archaeocyte – a pleuripotent or multipotent cell found in the mesohyle or a gemmule

Asconoid – sponge body plan where the entire spongocoel is lined by choanocytes

Atrium (spongocoel) – central cavity of the sponge body that carries water to the osculum

Boundary layer – thin film of water that lies directly on the surface of an object where there is no water flow; the thickness of the boundary layer is a function of water flow rate and shape and contour of the surface

Choanocyte – a cell bearing a single flagellum surrounded by a collar of microvilli, these are similar in structure and function to choanoflagellates

Choanocyte chamber – a space lined by flagellated chaonocytes

Egg – unfertilized female gamete

Embryo – the developmental stage between fertilization of an egg and hatching

Gap junction – connection between neighboring cells that makes the cytoplasm of neighboring cells continuous with each other and allows direct cell-to-cell communication

Gemmule – overwintering structure produced by some freshwater sponges, contains unspecialized thesocytes (amoebocyte cells) that produce a new sponge body after emerging

Incurrent canal – canals carrying water from ostia to choanocyte chambers

Leuconoid – the most complex sponge body plan, where only small spherical feeding chambers are lined by chaonocytes

Mesohyle – a gelatinous matrix located between the outer and inner body surfaces of sponges

Micropyle – opening in a gemmule through which thesocytes (binucleate archaeocytes) emerge

Multipotent – the ability of an undifferentiated cell to produce many kinds of specialized cells

Oscula (plural) / Osculum (singular) – excurrent opening(s) of a sponge

Ostia (plural) / Ostium (singular) – incurrent opening(s) of a sponge

Pleuripotent – the ability of an undifferentiated cell to produce many but not all kinds of cells

Prosopyle – openings through which water enters radial canals in the syconoid plan

Spicules – secreted $CaCO_3$ or SiO_2 structures that provide skeletal support and physical defense in sponges and protection to gemmules

Spongin – stiff yet flexible supporting material made of collagen, produced by many sponges of Class Demospongia

Spongocoel (atrium) – excurrent cavity where water collects prior to leaving a sponge body

Syconoid – sponge body plan where only radial canals are lined by choanocytes

Thesocyte – binucleate archaeocyte cells found in gemmules

Chapter 5: Phylum Cnidaria and Phylum Ctenophora

There are over 13,000 species in Phylum Cnidaria including jellyfish, corals, sea anemones and their relations. The cnidarian body plan is anatomically simple and includes swimming medusae and sessile polyps. Cnidarians contain true tissues with cells producing gap junctions. They also have a gut where external digestion takes place. Characteristics of Phylum Cnidaria are presented in **Table 5.1**.

Table 5.1. Characteristics of Phylum Cnidaria (after Brusca, *et al.* 2016).

- Diploblastic body with ectoderm (giving rise to epidermis), endoderm (giving rise to gastrodermis) and mesoglea (an ectodermally derived acellular gelatinous matrix between the epidermis and gastrodermis)
- Body symmetry is radial or biradial
- Cnidae/cnidocytes – cells that house stinging nematocysts
- Integrated neuromuscular system with myoepithelial (epitheliomuscular) cells and a diffuse nerve net (no central nervous system)
- Neurons are nonpolar and non-myelinated
- Life cycles include a planula larva, sessile polyp and free-swimming medusa stages or variations on this theme
- Gastrovascular cavity with only one opening

Cnidarians are carnivores and some also have symbiotic relationships with photosynthetic protists called zooxanthellae or zoochlorellae. Cnidarians use tentacles bearing batteries of cnidocytes with nematocysts for prey capture, defense and competition. Once captured, food is stuffed into the gastrovascular cavity where extracellular digestion takes place. A nematocyst can fire only once then a new one must replace it. Some cnidarians may use as many as 75% of their nematocysts per day.

A nematocyst shoots a hollow tube into its prey through which toxins are injected. Toxin potency varies greatly between species. Toxins of some species such as the west coast sea anemone *Anthopleura elegantissima* are so benign that you can run your exposed skin along their tentacles and feel nothing more than a little stickiness as if you were dragging a fingertip along a strip of sticky tape. You would however have a numb tongue for hours if you decided to lick one, as a student of mine once discovered. Other species of cnidarians have toxins so powerful that they are fatal to humans. The most notorious of these is the sea wasp box jelly *Chironex fleckeri*, a species that lives in the Indo-Pacific region.

Watch this video clip from the video series *The Shape of Life* for an introduction to the cnidarian body plan: http://shapeoflife.org/video/cnidarians-life-move

Phylum Cnidaria "nettle" – Taxonomy

Phylum Cnidaria has been arranged taxonomically many different ways over the years. The taxonomy used in this exercise is indicated below (after Brusca, *et al.*, 2016).

Subphylum and Class Anthozoa "flower animal" (included in this exercise). These are sea anemones, corals and their relations. Anthozoans lack the medusa stage in their life cycles. They have hollow tentacles that are extensions of the gastrovascular cavity and reproduction may be sexual, asexual or both.

Subphylum Medusozoa "jellyfish animal". These animals are cnidarians with polyp and medusae stages, and medusae may remain attached to polyps or be free-swimming. Gonads in this group are epidermal. There are four classes in this subphylum: Staurozoa, Cubozoa, Scyphozoa and Hydrozoa

> **Class Staurozoa "cross animal"** (not included in this exercise). These are stalked jellyfish and are unusual in that they produce non-ciliated planula larvae. These animals produce eight tentacle-bearing arms and are sessile, attached to the substrate via an adhesive stalk. Follow this link to see living stauromedusae: https://www.youtube.com/watch?v=YCB4sq_PAGQ

> **Class Cubozoa "box animal"** (not included in this exercise). These are box jellyfish. These animals produce planula larvae that develop into a tiny polyp stage that may or may not undergo asexual reproduction producing more polyps. Each polyp eventually produces a single medusa. Adult box jellies have a tall box-shaped bell with a more or less square base that bears a fleshly lip of tissue called the velarium that restricts the opening into the subumbrellar space. Each corner of the base of the bell bears a sensory rhopalium, a blade-like structure called a pedalium and either a single tentacle or a cluster of tentacles that bear nematocysts. Stings from these animals are particularly painful and can be fatal when medical help is not immediately available. Follow this link to learn more about box jellies: https://www.youtube.com/watch?v=WrMRwddl7iQ

> **Class Scyphozoa "cup animal"** (included in this exercise). These are the true jellyfish. Scyphozoan planulae develop into small polyps that produce many medusae asexually via strobilation. The free-swimming medusae stage is sexual and dominates the life cycle. These medusae lack a velum and the margin of the bell is often divided into shallow scalloped edges called lappets. Sensory structures called rhopalia are located in the notches between lappets. Follow this link to see swimming *Aurelia* (moon jellyfish), the same species highlighted in this exercise: https://www.youtube.com/watch?v=oNgeR_xFwpQ

> **Class Hydrozoa "water animal"** (included in this exercise). This group includes small colony-forming animals called hydrozoans or hydroids and solitary *Hydra*. Polyp and medusa stages may both be prominent in these life cycles. The hydrozoan planula develops into a sessile polyp that undergoes cloning to produce a colony of zooids that may or may not exhibit polymorphism. Some zooids produce hydromedusae by budding. Medusae may or may not be released to a free-swimming existence. Tentacles may be solid or hollow. Hydromedusae are relatively small and have a tall bell with a restrictive velum surrounding the bell opening. There are usually four radial canals plus a ring canal that runs around the edge of the bell in hydromedusae.

Subphylum Myxozoa "mucus animal" (not included in this exercise). These are microscopic cnidarians with a convoluted taxonomic history of being included in several different other groups, though recent molecular and anatomical analysis confirm that this is a group of highly derived cnidarians. Myxozoans are intracellular parasites of vertebrate and invertebrate hosts; they have unusual shell-like valves and polar capsules bearing nematocyst-like structures. Follow this link to find out why these animals are now included in Phylum Cnidaria: https://www.youtube.com/watch?v=Me0zdgFVrhM

Subphylum (and Class) Anthozoa – sea anemones, corals and relations

The anthozoan life cycle includes only gonochoric (sexually reproductive) polyps that in many species also undergo clonal growth by budding, binary fission or pedal laceration. Anthozoan polyps are large compared to medusozoan and hydrozoan polyps. Anthozoans usually require hard substrates to survive but a few species have adapted to soft-sediment habitats.

Tasks – Subphylum Anthozoa

Metridium – plumose or frilled sea anemone

1) Immerse a preserved specimen of *Metridium*. These animals are beautiful when they are alive; the body column is cream to white in color and the finely branched tentacles can be brilliant white but sadly become a brownish gray when preserved. DRAW your specimen. The size of anemone tentacles provides a clue about their lifestyle; finely branched tentacles are used to capture tiny plankton and larger tentacles are used to capture bigger prey including fishes. Based on this premise, how do you think *Metridium* makes a living? Use **Fig. 5.1** to help you identify what you see.

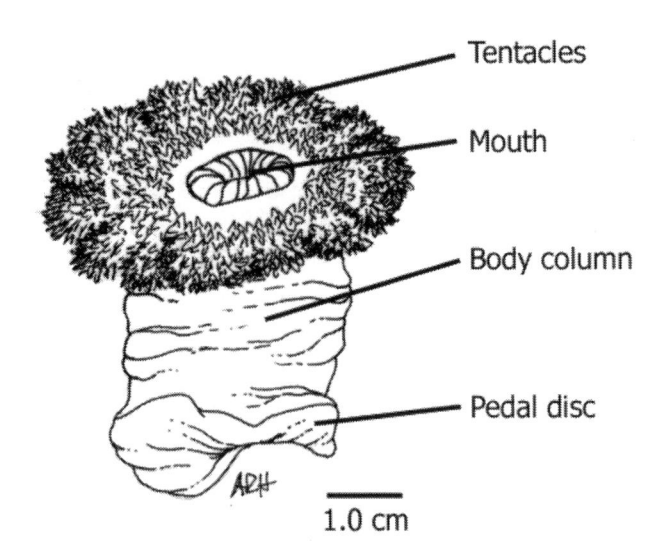

Figure 5.1. External anatomy of *Metridium*. (Image: ARH)

2) Carry out the following dissections with a lab partner. DRAW what you see as you carry out these dissections and expose new structures.
    a. **Dissection #1**: Internal anatomy, longitudinal view. Use scissors, scalpel or razor blade to make a longitudinal cut that runs the length of the body through the

mouth and pharynx all the way through the pedal disc and produces mirror image halves of your specimen. After you complete the initial cut, each of you can take one half of the animal for individual examination. Look for the mouth, pharynx and gastrovascular space. Once you have found these look for complete septae that run between the pharynx and the body wall and incomplete septae that extend inward from the body wall into the gastrovascular cavity but are not attached to the pharyngeal wall. DRAW what you see and use **Fig. 5.2** to help you identify the anatomy of your specimen.

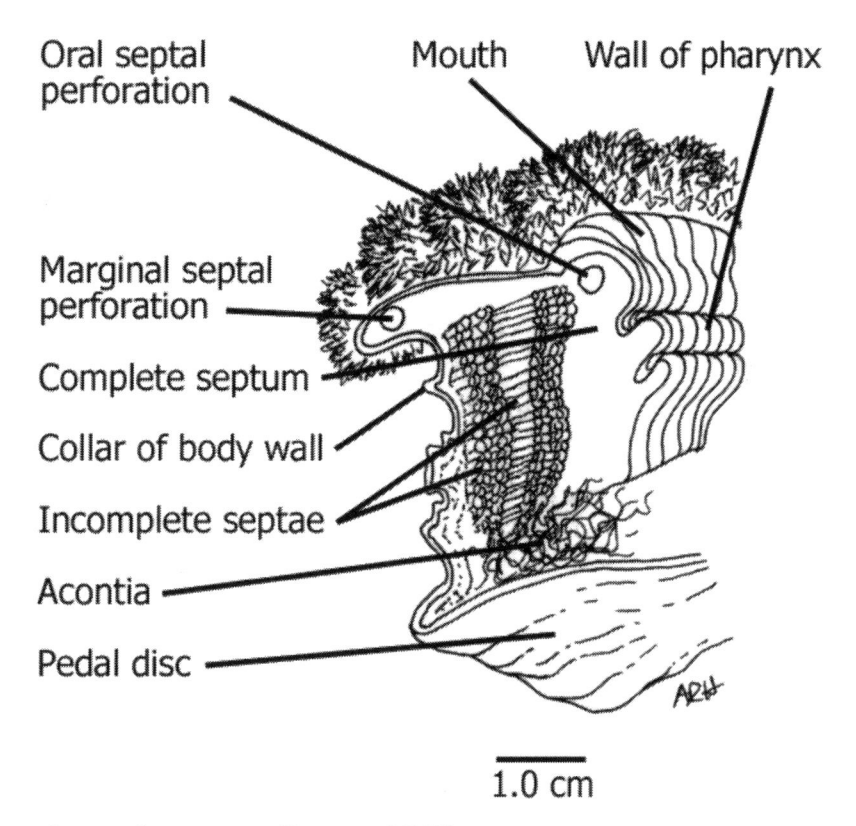

Figure 5.2. *Metridium*, internal anatomy. (Image: ARH)

 b. **Dissection #2**: Internal anatomy, transverse view. Obtain another preserved specimen. Use a scalpel or razor blade to make two or three transverse cuts through the body column perpendicular to the central axis of the body. One cut should pass through the pharynx and another should pass through the gastrovascular cavity at the base of the body column. DRAW each transverse section and use **Fig. 5.3** to help you identify the anatomy of your specimen.

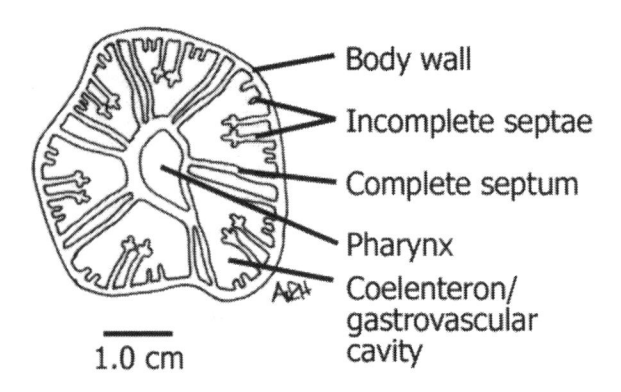

Figure 5.3. Transverse view through the pharynx of *Metridium*. (Image: ARH)

Coral

Corals are prominent members of Anthozoa in some marine communities. They are perhaps best known from tropical coral reef systems but corals also produce deep cold-water reef communities and soft-corals are prominent Antarctic benthic communities. Reef-building corals are referred to as hermatypic corals.

3) Peruse the diversity of coral specimens provided and choose one specimen to study in detail. Look first at the overall shape of the coral colony and consider advantages and disadvantages of that colony form. WRITE your observations and thoughts about the coral of your choice in your notebook. Use **Fig. 5.4** to help you identify what you see. The coral skeleton shown in **Fig. 5.4** has particularly large calyces. You may need to use a magnifying lens or dissection scope to examine the structure of a calyx in the coral you are investigating. Only the skeleton of the coral animal is present, no soft tissue remains. DRAW what you see.

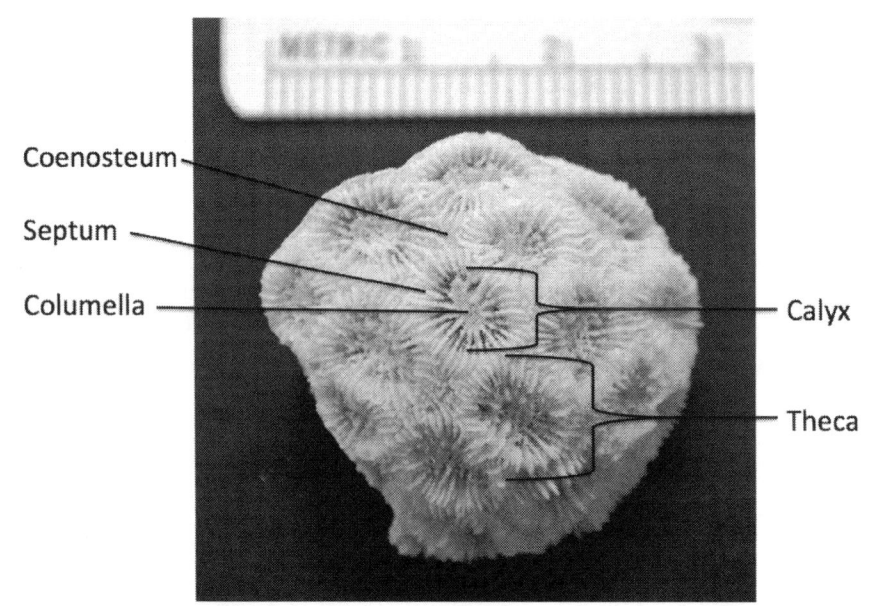

Figure 5.4. Skeleton of the Caribbean rough star coral *Isophyllastrea rigidia*. (Image: ARH)

Subphylum Medusozoa

Class Scyphozoa – true jellyfish

Scyphozoans have a gonochoric medusa stage that dominates the life cycle and many species have small sessile polyps that produce medusae via strobilation, a form of cloning.

Tasks – Class Scyphozoa

Aurelia – moon jellyfish

1) Immerse a preserved adult medusa of *Aurelia* and DRAW it. Most of the volume of jellyfish is made up of the thick layer of mesoglea located between the epidermis and gastrodermis of the bell. Refer to **Fig. 5.5** to help you identify what you see.

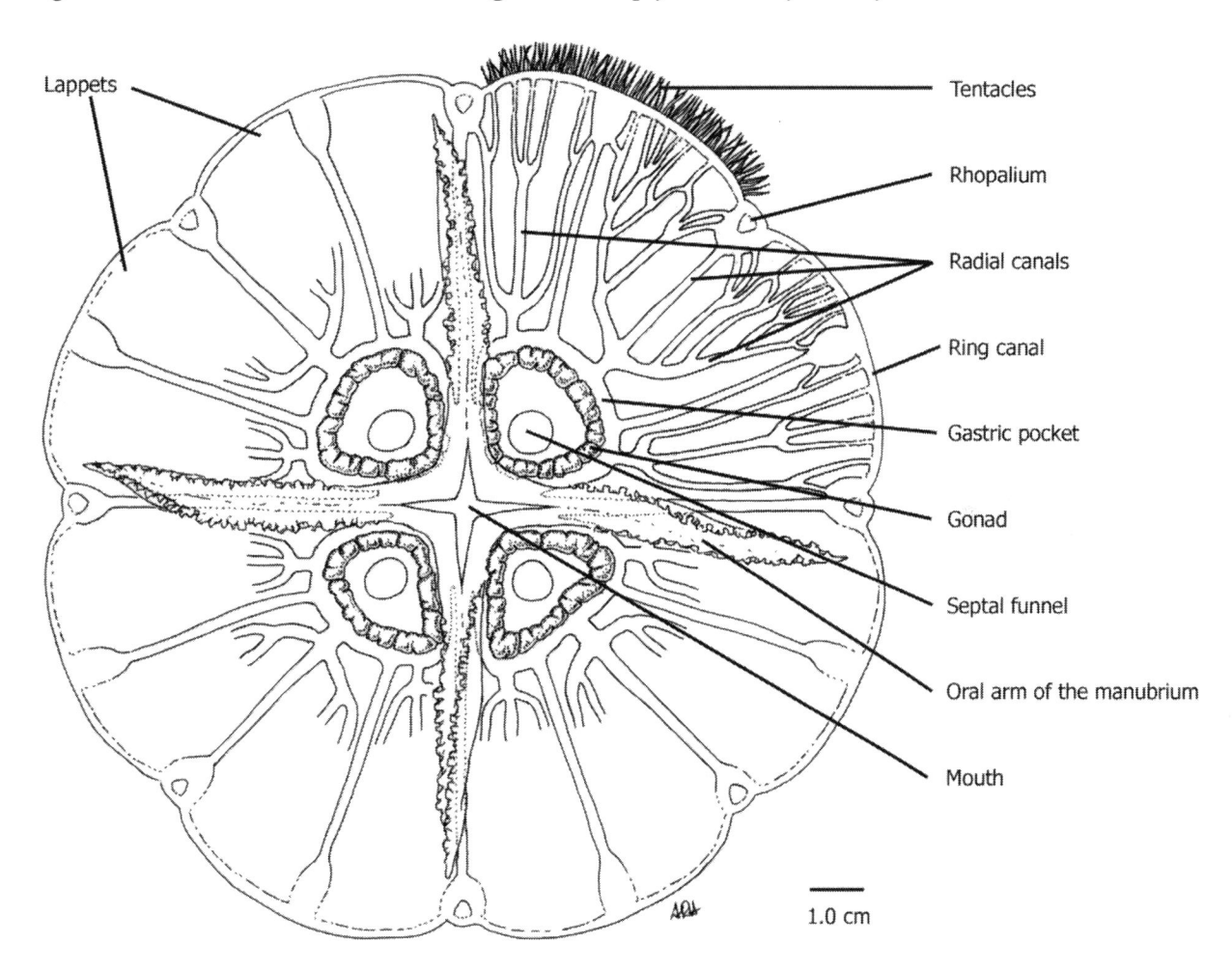

Figure 5.5. Anatomy of *Aurelia* medusa, subumbrellar view. (Image: ARH)

2) DRAW the following life history stages of *Aurelia* from prepared slides: planula, scyphistoma, strobila and ephyra. The arrows indicate the order of development from planula to ephyrae. Use **Fig. 5.6** to help you identify what you see.

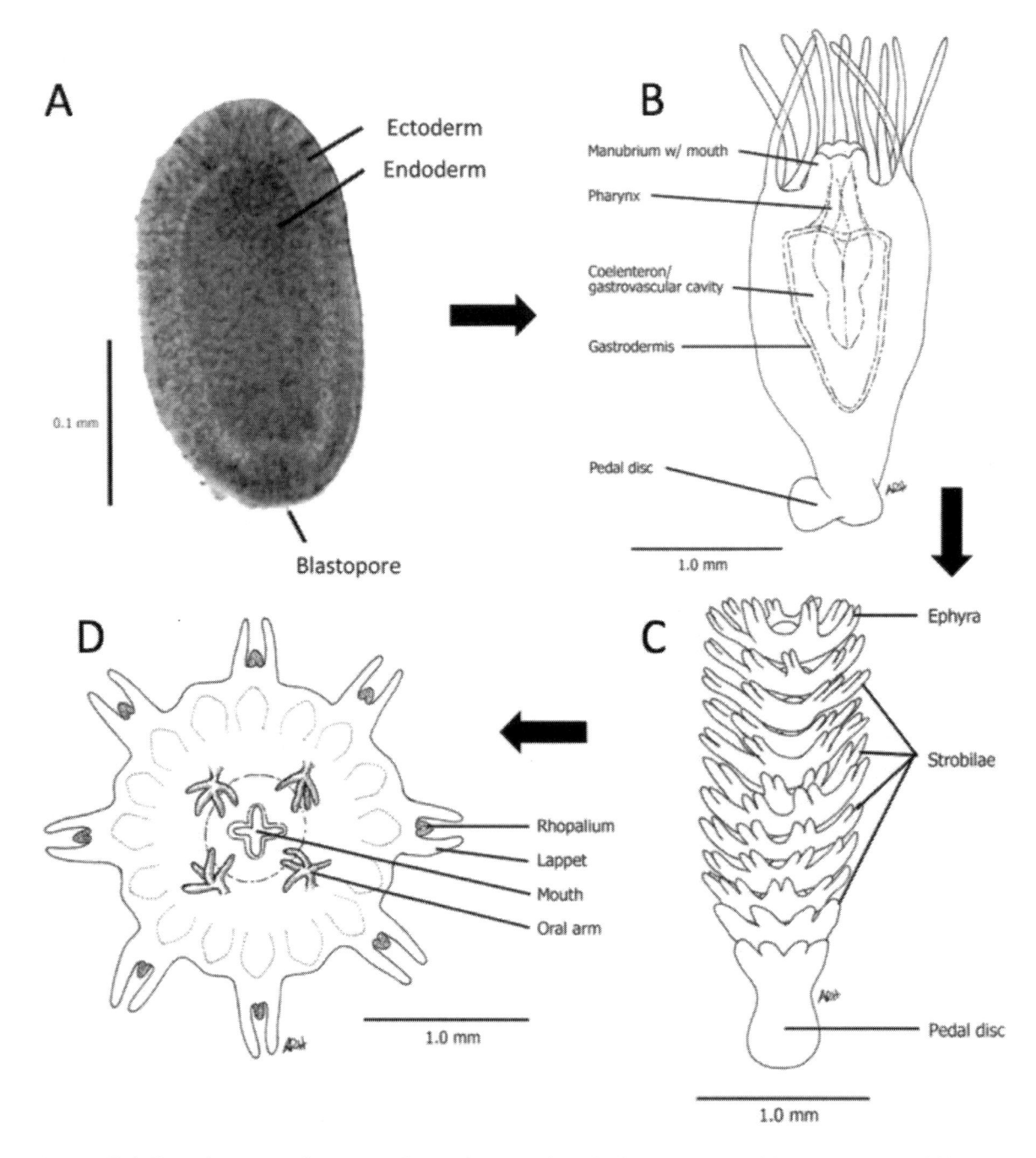

Figure 5.6. Developmental stages of *Aurelia:* **1)** planula larva, **2)** scyphistoma, **3)** strobila, and **4)** ephyra (immature medusa). Arrows indicate the order of these life stages in the life cycle. (Image: ARH)

Class Hydrozoa – hydroids and relations

Most members of this group are relatively small and their functional units are called zooids. The hydrozoan life cycle typically includes a clonal colony-forming polyps and gonochoric hydromedusae.

Tasks – Class Hydrozoa

The Portuguese man o' war *Physalia* is a free-floating colony belonging to Order Siphonophora, Class Hydrozoa. These colonies have heteromorphic zooids that cooperate to meet the colony's needs. The large gas bladder, the pneumatophore, is used for flotation, clusters of gastrozooids carry out digestion, dactylozooids are used for prey capture and defense and gonozooids are the reproductive members of the colony. A complete set of zooids, one of each type, is added as a group called a cormidium as the colony grows.

1) Observe a preserved Portuguese man o' war. Consider the advantages and disadvantages of a floating colony versus a sessile colony and WRITE your observations and thoughts in your lab notebook. Refer to **Fig. 5.7** to help you identify what you see. Small bulbous gonozooids are not visible in this photograph.

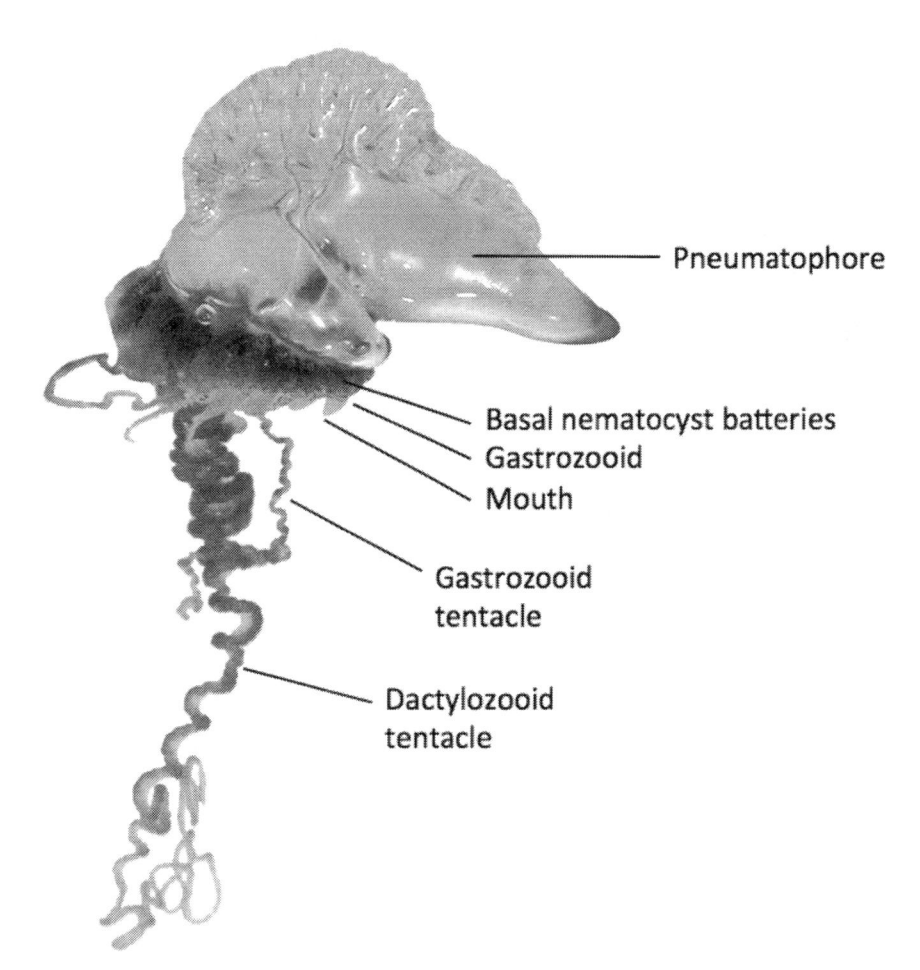

Figure 5.7. The Portuguese man o' war *Physalia physalis*. (Image: ARH, modified from image courtesy of NOAA)

Hydra is a common hydrozoan in freshwater lakes and ponds. It has an atypical life history for a hydrozoan because it lacks a medusa stage and lives as physiologically independent polyps. *Hydra* does however carry out sexual and clonal reproduction. *Hydra* produces planula larvae that develop into the polyp stage. Once a polyp grows large enough it carries out cloning via budding. Polyps reproduce sexually by producing ovaries or testes. It is nonetheless a good example of the hydrozoan polyp body plan with epidermal, mesogleal and gastrodermal layers.

2) Use a magnifying glass or dissection microscope to observe live *Hydra*. DRAW its body form and describe its behavior. Add some *Daphnia* or other small zooplankton common in freshwater to the sample and observe any changes in *Hydra* behavior. WATCH this short video if live specimens are not available:
https://www.youtube.com/watch?v=TfaafxnnJlY

3) Observe a prepared cross-section slide of *Hydra* and DRAW the three-layered body wall containing epidermis, mesoglea and gastrodermis. Use **Fig. 5.8** to help you identify what you see.

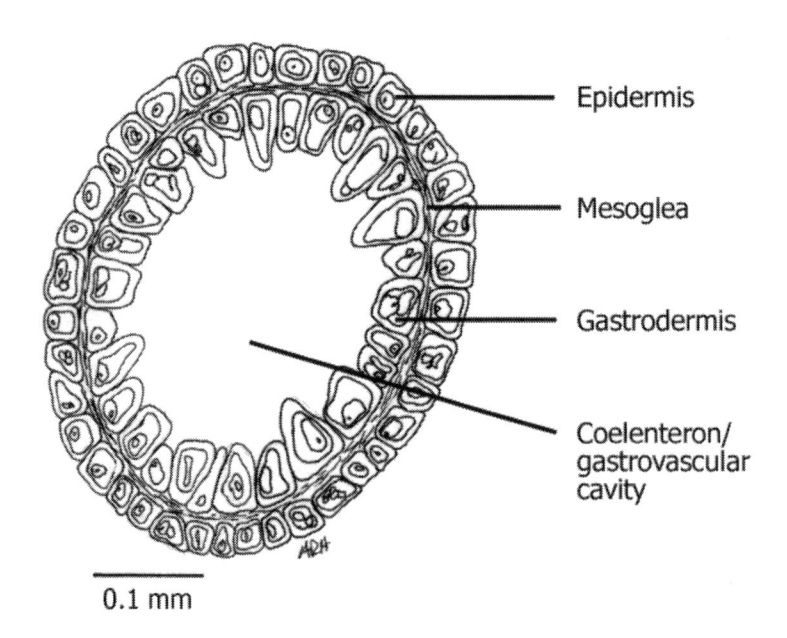

Figure 5.8. Cross-section through the body column of *Hydra*. (Image: ARH)

4) Observe prepared slides of budding, and male and female *Hydra*. DRAW a composite figure that shows budding, female and male reproductive structures in one image. Use **Fig. 5.9** to help you identify these structures. Sexually reproductive individuals can have multiple ovaries or testes.

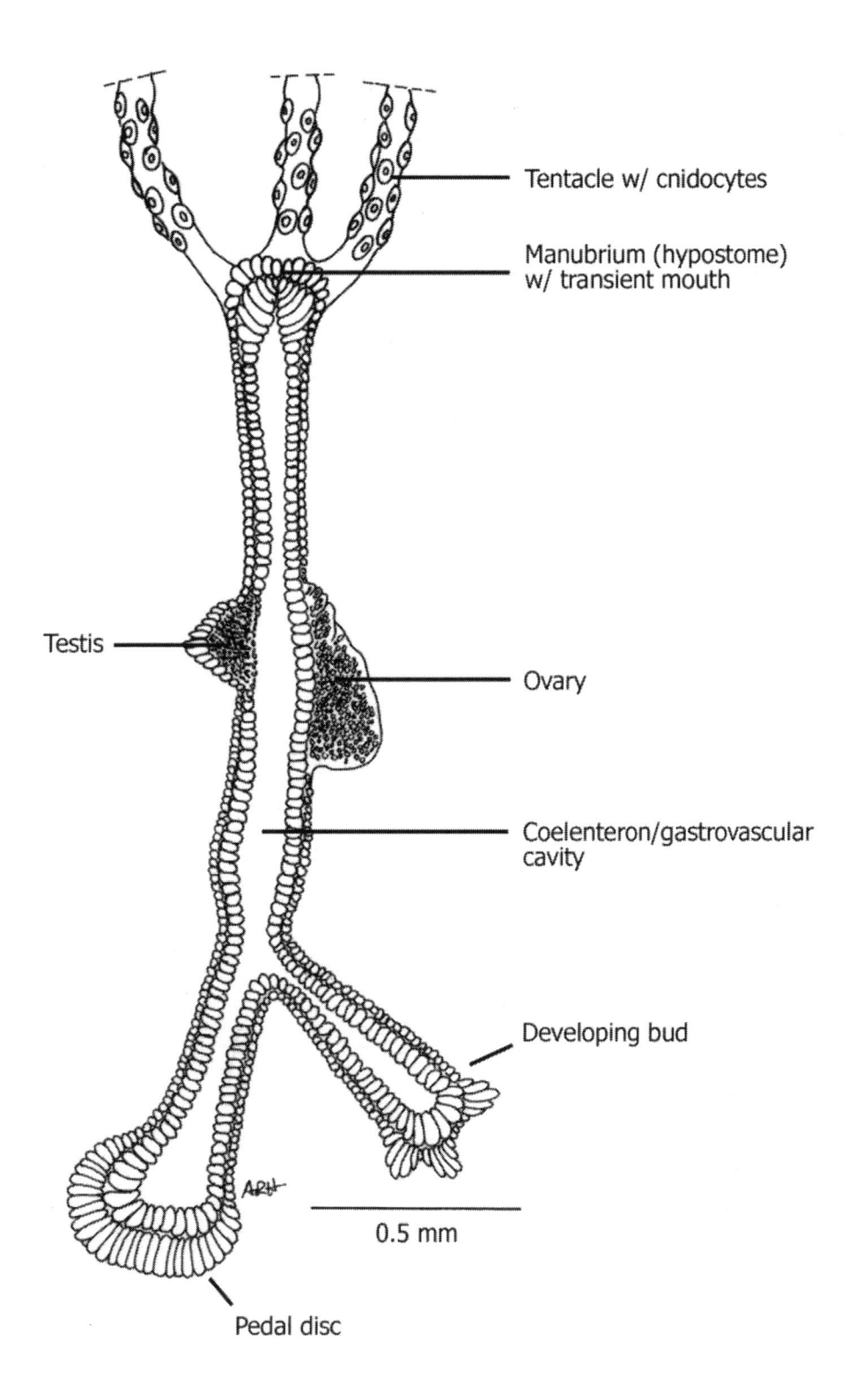

Figure 5.9. Composite drawing of a longitudinal section of *Hydra* including a bud, ovary and testis. (Image: ARH)

5) Observe a prepared slide of *Hydra* stained to show cnidocytes on a tentacle. Use **Fig. 5.10** to help you identify what you see.

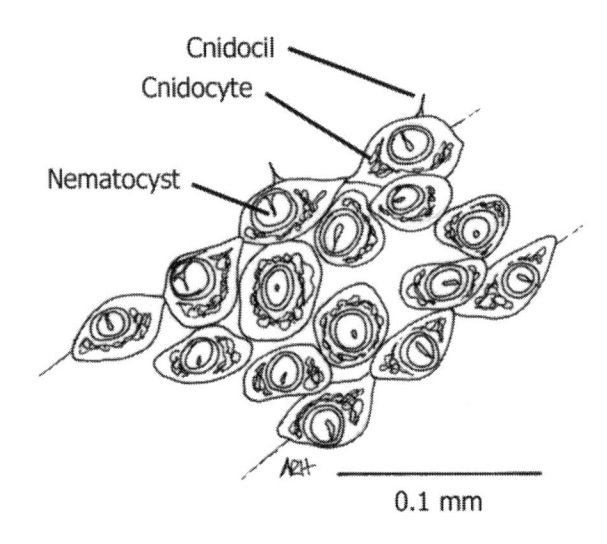

Figure 5.10. Cnidocytes on a tentacle of *Hydra*. (Image: ARH)

Obelia

Obelia is a commonly studied marine hydrozoan. It is more representative of the hydrozoan body plan and life cycle than *Hydra*. *Obelia* polyps grow as clonal polymorphic colonies. The medusa of *Obelia* is gonochoric (sexual).

1) DRAW an *Obelia* medusa from a preserved specimen or prepared slide. Use **Fig. 5.11** to help you identify what you see.

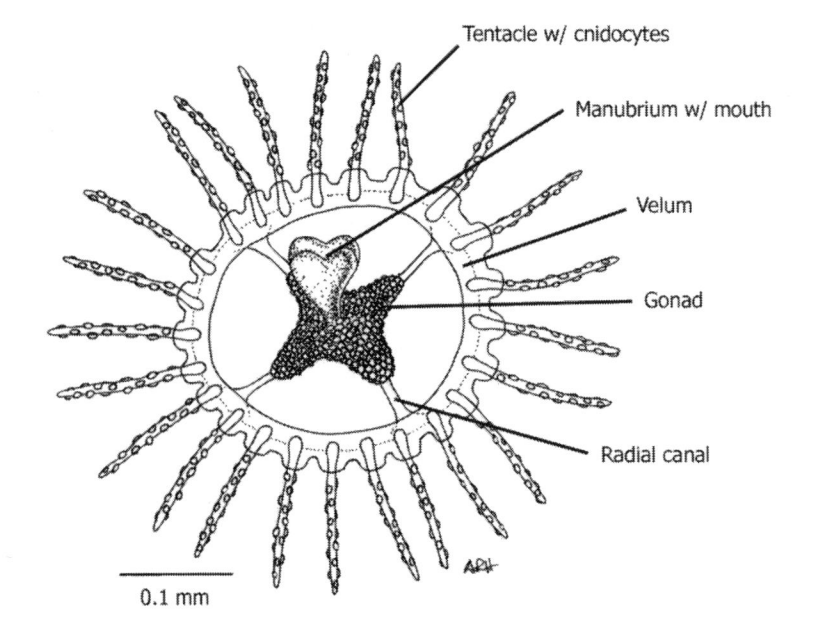

Figure 5.11. Anatomy of *Obelia* hydromedusa, subumbrellar view. (Image: ARH)

2) Observe either a prepared slide or a preserved specimen of an *Obelia* colony. DRAW and label a portion of the colony. Include at least one gastrozooid and one gonozooid in your drawing. Use **Fig. 5.12** to help you identify what you see.

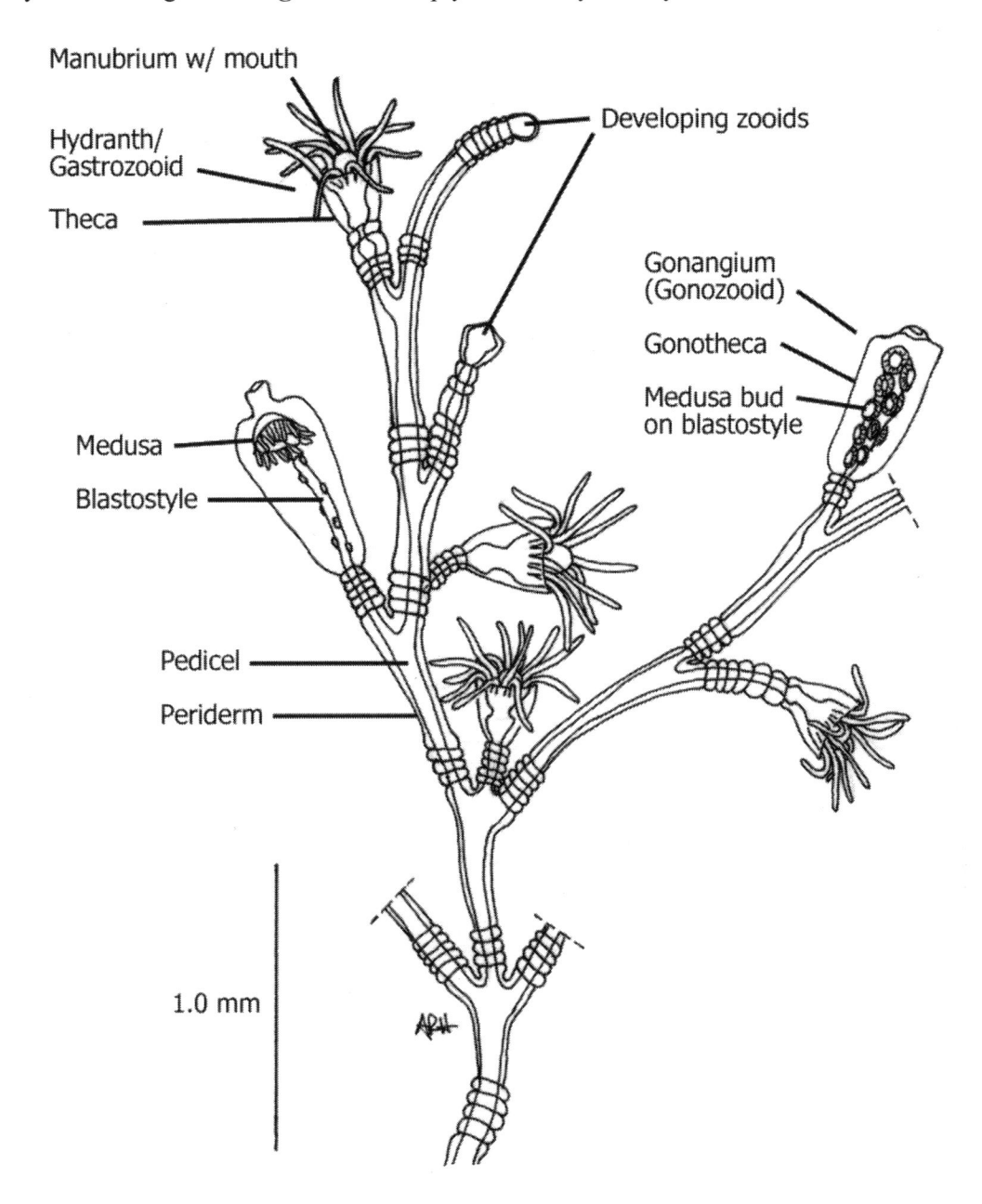

Figure 5.12. *Obelia* colony (polyp stage). (Image: ARH)

Phylum Ctenophora "comb bearing" - ctenophores

Ctenophores are the largest animals that move solely via ciliary action. They are strictly planktonic and like cnidarians are carnivores. They prey on a wide variety of plankton, including each other. These animals use sticky colloblast cells, lobe-like appendages or large mouths to capture and ingest prey. Characteristics of Phylum Ctenophora are listed in **Table 5.2**.

These videos show swimming and feeding behavior of ctenophores:
https://www.youtube.com/watch?v=zsMUeo4qJjk
https://www.youtube.com/watch?v=MmoChWQ6xCk

Table 5.2. Characteristics of Phylum Ctenophora (after Brusca, *et al.* 2016).

- Diploblastic, possibly triploblastic animals
- Biradial symmetry
- Colloblasts
- Branching gastrovascular cavity
- Eight rows of ciliary plates called ctenes

Tasks - *Pleurobrachia* – sea gooseberry or sea walnut
1) Examine the anatomy of *Pleurobrachia*. DRAW what you see and use **Fig. 5.13** to help you identify structures on your specimen. Note: You will probably not see the tentacles extended; they were most likely retracted when the animal was preserved.

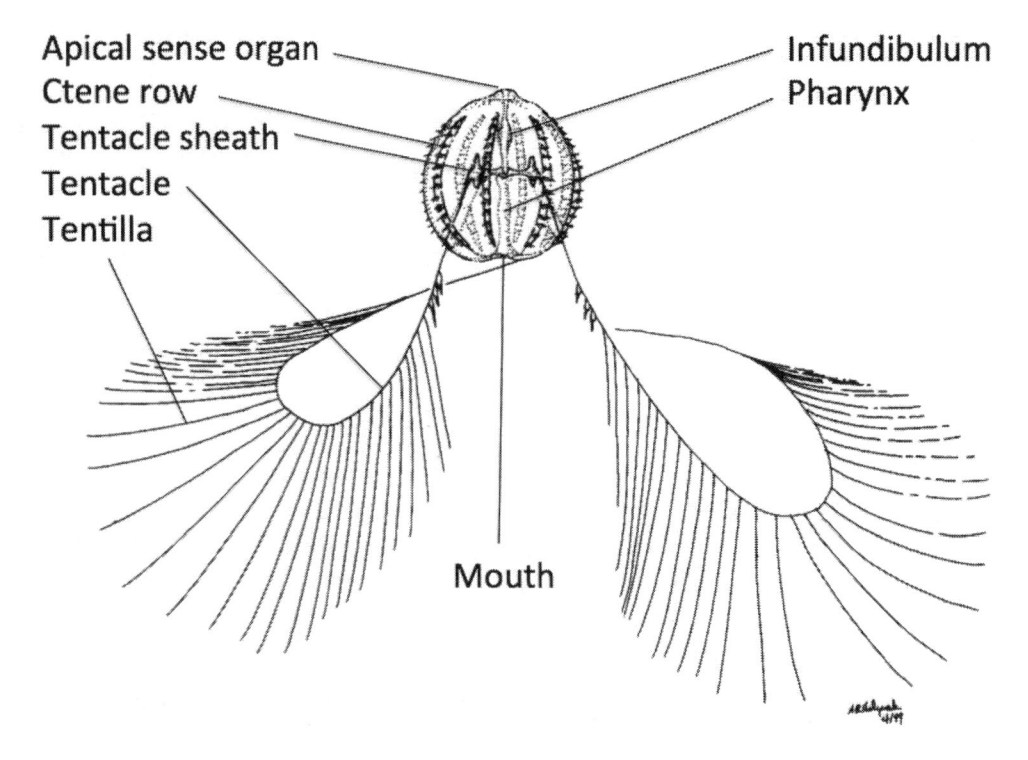

Figure 5.13. The ctenophore *Pleurobrachia* with tentacles and tentilla extended. (Image: ARH)

Group Questions

1) Cnidarians were probably the first hunters and have been around for a long, long time. They continue to be extremely successful both ecologically and evolutionarily. What attributes of cnidarians do you think help explain their ongoing success?
2) Develop a set of hypotheses that explain when it would be more advantageous to a cnidarian that is colonial and when it would be more advantageous to be solitary.
3) What are costs and benefits of radial symmetry?

Phylum Cnidaria – Glossary

Acontia – cnidae-bearing filaments that are extensions of internal septae in anthozoans, used for digestion and feeding and can be extended though openings in the body wall for defense

Astogeny – size, shape and pattern of development of a colony

Binary fission – clonal growth that takes place when the entire body divides into two equal halves and each half produces missing structures

Biradial symmetry – body plan that can be divided into equal halves in only two planes, usually because the mouth and pharynx are oblong within an otherwise radially symmetrical body

Blastopore – opening of the archenteron, produced during gastrulation

Blastostyle – rod-like structure in the center of a hydrozoan gonozooid where medusa buds are produced

Body column – cylindrical body of an anemone

Budding – cloning that occurs when a mass of cells pinches off of the parent body and the resulting bud develops into a genetically identical but physiologically independent individual

Calyx – cup-like depression that houses soft tissue of a coral polyp

Cnidocil – trigger-like structure that in many cases must be touched to fire a nematocyst

Cnidocyte (cnida) – cell that produces a stinging nematocyst

Coelenteron/gastrovascular cavity – fluid-filled space within the body where digestion and gas exchange takes place

Coenosteum – skeletal material between thecae of coral polyps

Collar of body wall – sphincter muscle on the body column just below the tentacles of anemones that constricts when the tentacles are retracted

Columella – center of the calyx of a coral skeleton

Complete septum – sheet of tissue extending from the body wall to the inner wall of the pharynx dividing the gastrovascular cavity into multiple compartments in anthozoans

Dactylozooid – zooid that carries out defense and prey capture in the Portuguese man o' war

Diploblastic – body plan that produces only ectoderm and endoderm embryonic tissues

Ectoderm – embryonic tissue layer that covers the exterior of the embryo and gives rise to the epidermis

Endoderm – embryonic tissue layer that comprises the archenteron and gives rise to the gastrodermis in cnidarians

Ephyra – immature free-swimming scyphomedusa stage produced by cloning

Epidermis – tissue derived from embryonic ectoderm that covers the outer surface of the body

Epitheliomuscular cell – muscle cell that has an elongate contractile element and an extension of the cell body that is exposed to the external environment or gastrovascular cavity

Gastric pocket (pouch) – invagination of the epidermis that houses the gonads and is the location of extracellular digestion in scyphomedusae

Gastrodermis – tissue derived from the embryonic endoderm that lines the gastrovascular cavity

Gastrozooid – zooid that carries out digestion of captured prey in the Portuguese man o' war

Gonangium (gonozooid) – hydrozoan zooids that produce hydromedusae by budding

Gonochoric – having separate female and male individuals

Gonotheca – chitinous covering that protects gonozooids

Gonozooid – zooid that carries out reproduction in the Portuguese man o' war, also a structure that produces medusa buds in *Obelia* by cloning

Hydranth/gastrozooid – feeding zooids of hydrozoans

Hypostome – dome-shaped structure in the center of a ring of tentacles in hydrozoans that bears the mouth

Incomplete septum – Sheets of tissue in anthozoans that extend from the inner body wall partway into the gastrovascular cavity but do not reach the pharynx, they increase surface area for absorption, and margins of these septae are trilobate in cross-section and bear cnidae and secrete digestive enzymes

Lappet – scallop-shaped margin of the scyphomedusa bell

Manubrium (with mouth) – tube of tissue bearing the mouth in some cnidarians

Marginal septal perforation (ostium) – openings through the distal edges of all complete septae in anthozoans, they provide communication between neighboring compartments

Medusa – free-swimming life stage that carries out sexual reproduction

Medusa bud – small button-shaped mass of tissue that pinches off and swims away as an immature hydromedusa in hydrozoans

Mesoglea – gelatinous matrix of material derived from the ectoderm that forms a layer between the epidermis and gastrodermis

Mouth – only opening into a gastrovascular cavity, functions as both mouth and anus

Nematocyst – stinging organelle that contains a coiled hollow tube that everts when it fires, penetrating its target and injecting toxins into it

Oral arm of the manubrium – tapering elongate structures attached to the short manubrium of scyphomedusae, food is captured by the short frilly tentacles of the arms and is moved to the mouth in grooves that run along the aboral surface of each arm

Oral disc – flat area surrounding the mouth of anemones that is free of tentacles

Oral septal perforation (ostium) – medial openings through complete septae of anthozoans, provides communication between neighboring compartments of the gastrovascular cavity

Pedal disc – surface of attachment to the substrate, anemones can use the pedal disc to crawl

Pedal laceration – form of clonal growth where small bits of the pedal disc are torn off and left behind as an animal moves, each of these small masses of tissue develops into a new individual

Pedicel/coenosarc – tissue that connects members of a hydrozoan colony to each other

Periderm – transparent chitinous covering that surrounds and protects the pedicel

Pharynx – passageway that moves food from the mouth to the gastrovascular cavity

Planula – ciliated larval stage of cnidarians

Pneumatophore – specialized zooid that is the float or gas sac of the Portuguese man o' war

Radial canal – extension of the gastrovascular cavity that carries nutrients and waste materials in scyphomedusae

Radial symmetry – body plan with a central axis that can be divided into many different mirror image halves

Rhopalium – a sensory organ located between lappets in scyphomedusae, it includes chemosensory cells, a statocyst (senses gravity) and an ocellus (senses light)

Ring canal – canal that runs around the margin of the bell of a medusa

Scyphistoma – small feeding polyp stage of scyphozoans

Septae (coral) – thin vertical ridges within the calyx of a coral

Septal funnel – an opening into the gastric pouch from the gastrovascular cavity in scyphomedusae

Siphonoglyph – a ciliated band of tissue in the pharynx between the mouth and gastrovascular cavity of anthozoans, re-inflates the gastrovascular cavity after it has deflated

Strobila – polyp stage that develops from the scyphistoma stage in scyphozoans and produces ephyrae via strobilation

Tentacle – cnidae-bearing structures used for feeding, the size of tentacle reflects prey size

Tentacles (marginal) – Thin sometimes very long tentacles that produce a fringe on lappets, their main function is to generate eddies that pull particles toward the mouth in scyphomedusae

Theca – transparent chitinous covering of zooids in hydrozoans

Velum – inward-facing shelf of tissue constricting the subumbrellar opening in hydromedusae, allows the animal to produce an increased velocity of water flow each time the bell contracts

Phylum Ctenophora – Glossary

Apical sense organ – structure containing a statocyst that regulates beating of the ctene rows, also includes ciliated sensory fields presumably for monitoring water quality

Ctene – A plate of 100s of cilia fused together and beating together so they function more like an oar than a whip

Ctene row – a row of many ctenes that beat synchronously to provide propulsion

Infundibulum – a hollow structure that connects the aboral sense organ to other parts of the body

Pharynx – muscular tube that moves food from the mouth to the branched gastrovascular cavity

Tentacle – primary filamentous extension of the body, bears many smaller tentilla

Tentilla – extremely fine tentacle-like structures that extend off of a tentacle, these bear sticky colloblasts and are the main food capturing structures in tentacle-bearing ctenophores

Tentacle sheath – cavity into which the tentacles can be withdrawn entirely, these sheaths are not connected to the digestive tract

Chapter 6: Phylum Platyhelminthes

Phylum Platyhelminthes is the first taxon in this manual that represents Clade Bilateria, bilaterally symmetrical animals. Clade Bilateria includes the Clades Spiralia and Ecdysozoa. Spiralians originally included only animals that exhibited spiral cleavage during early development, but this clade now includes animals that are bilateral and use cilia for locomotion or feeding and that share molecular traits not discussed here. Clade Ecdysozoa includes animals that produce a protective outer cuticle or exoskeleton that in most cases is molted via a specific process called ecdysis. We will return to ecdysozoans later in this lab manual.

As you work through this laboratory exercise ponder on advantages and disadvantages of a bilaterally symmetrical body plan as opposed to a radial or asymmetrical plan. The bilateral plan is obviously an extremely successful one since this is what most animals have.

There are about 27,000 described species of living Platyhelminthes, the flatworms. This group includes free-living worms and medically important parasites. These animals do not exhibit any single unique feature or trait so we use a combination of traits to define the body plan of this phylum. Characteristics of Platyhelminthes are listed in **Table 6.1**.

Table 6.1. Characteristics of Phylum Platyhelminthes (after Brusca, *et al.* 2016).

- Unsegmented worm-shaped body
- Triploblastic construction
- Bilateral symmetry
- Cephalization and a central nervous system with a cephalic ganglion and longitudinal nerve chords
- Acoelomate body
- Protonephridia or other specialized excretory structures
- Spiral cleavage with mesoderm derived from the 4d cell
- Body is dorso-ventrally flattened

Phylum Platyhelminthes includes a wide diversity of body plans and life styles. There are free-living freshwater, marine and terrestrial flatworms that are largely opportunistic chemosensory hunters and scavengers. It is the medically important parasitic flukes and tapeworms however that occupies the vast majority of scientific attention. About three quarters of all described flatworms are parasites. Follow this link and watch this video introduction to the flatworms from the *Shape of Life* video series: http://shapeoflife.org/video/flatworms-first-hunter.

Phylum Platyhelminthes "flat worm" – Taxonomy

A detailed description of the taxonomy of flatworms cannot adequately be covered here so only a few prominent groups are mentioned.

Order Polycladida "many branches" (not covered in this exercise) – Polyclad flatworms. This is a group of marine species that have a gastrovascular cavity with many diverticulae (out pockets or side branches). The largest of these worms can be up to 15 cm long and are quite colorful though most are smaller than 3 cm. Some polyclads can swim with an elegant undulation of the margins of the flattened body. Follow this link to see a swimming polyclad flatworm: https://www.youtube.com/watch?v=eNZS8Th8GYo

Order Tricladida "three branches" (covered in this exercise) – Triclad flatworms. This group contains worms that live in marine, freshwater and terrestrial habitats. These worms all have three main branches of the gastrovascular cavity, two extending posteriorly and one anteriorly from the middle of the body. Most students of zoology usually learn about planarian triclad worms but if you follow this link you will see some footage of a terrestrial worm called the arrowhead or hammerhead flatworm (see also **Fig. 6.1)**: https://www.youtube.com/watch?v=-oSGM2pNl4I

Terrestrial triclad flatworms are not particularly rare where it is moist and warm. **Figure 6.1** shows a terrestrial arrowhead flatworm seen crawling on an exterior window in western Arkansas, USA. These worms are harmless to humans; earthworms are their main prey. We now know that at least some terrestrial triclads produce tetrodotoxin, the first report of this class of toxins in any terrestrial invertebrate (Stokes, *et al*, 2014 - doi:10.1371/journal.pone.0100718)

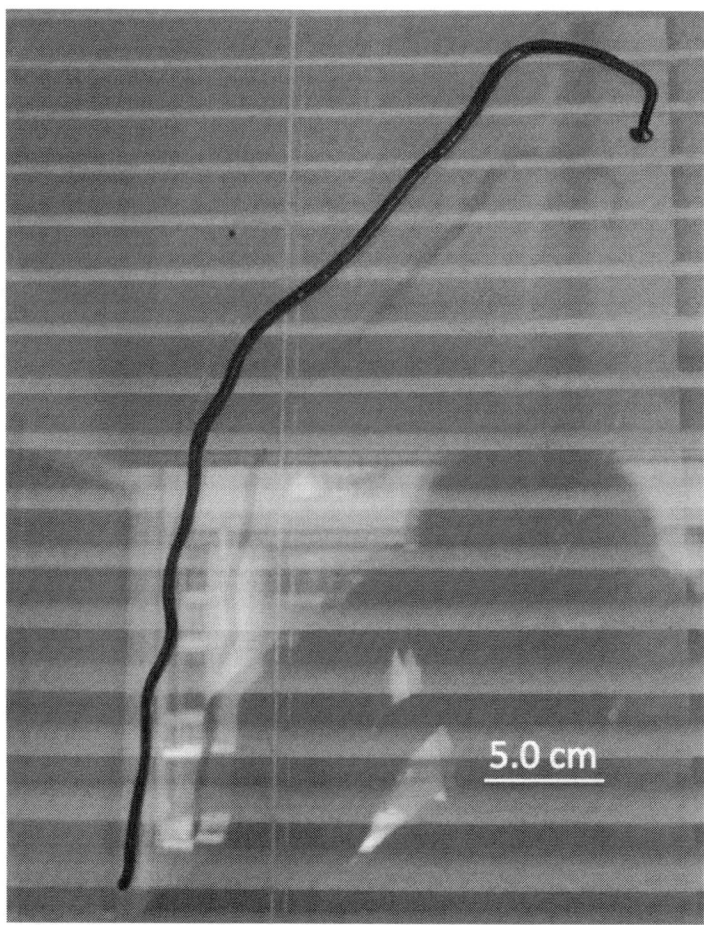

Figure 6.1. Arrowhead flatworm, Arkansas, USA. (Image: Courtesy of Ryan Durrant)

Infraclass Neodermata "new skin" – digenean and monogenean flukes and tapeworms. These worms produce a ciliated epidermis when they are larvae, but a syncytial epidermis replaces the cellular epidermis by the time they become adults. Members of this group are exclusively parasitic.

> **Cohort Trematoda "fluke"** (covered in this exercise) – endoparasitic digenean flukes. All of the members of this group have complex parasitic life histories, i.e., they have life histories that include at least two hosts. The intermediate hosts in these life cycles are typically molluscs and the final hosts are vertebrates. The infective stages of flukes can change the behavior of their intermediate hosts in order to increase the likelihood of the parasite finding a suitable final host. This video shows how flukes can change host behavior:
> https://www.youtube.com/watch?v=Go_LIz7kTok

> **Cohort Monogenea "one generation"** (covered in this exercise) – ectoparasitic monogenean flukes. All of the members of this group have simple ectoparasitic life histories. They live attached to the external surface of their host and they have only one host in the life cycle. These animals locate a suitable host when they are larvae and then metamorphose into adults. Hosts of freshwater monogeneans include primarily fishes plus a few kinds of other kinds of vertebrates, and cephalopod molluscs for marine species. It is sometimes possible to find monogeneans on fish purchased from big box stores or fishes that live in small warm freshwater lakes and ponds. Monogeneans are most commonly found attached to the gills and fin surfaces of fishes. Follow this link to see monogenean flukes attached to fish gill tissue – there are two species shown, a smaller species you'll see at the beginning of the video and a larger one the video spends more time focusing on toward the end.
> https://www.youtube.com/watch?v=OiOeznXQZuI

> **Cohort Cestoda "girdle"** (covered in this exercise) – the tapeworms. Tapeworms are highly derived, so much so that they do not have mouths or guts. Instead their outer body covering is adapted to absorb predigested materials from the intestines of their hosts. These worms have a scolex that bears suckers, hooks or both and is used to attach the worm to a host's intestinal wall. The rest of the body is comprised of structures called proglottids, each of which bears both female and male reproductive organs and can produce huge numbers of offspring. Tapeworms, like digenean flukes, have complex life histories though intermediate and definitive hosts of tapeworms are usually both vertebrates.

<u>Tasks – Order Tricladida – triclad flatworms</u>

1) Observe a live planarian. Use a magnifying lens or dissection scope to describe its body form and behavior. WRITE your observations in your lab notebook.
2) Work with a partner to carry out a simple experiment to see whether planarians prefer light or dark conditions. WRITE your hypotheses, methods, results and conclusions in your lab notebook.
3) DRAW a prepared whole mount slide of a planarian stained specifically to highlight the digestive tract. Refer to **Fig. 6.2** to help you identify what you see.

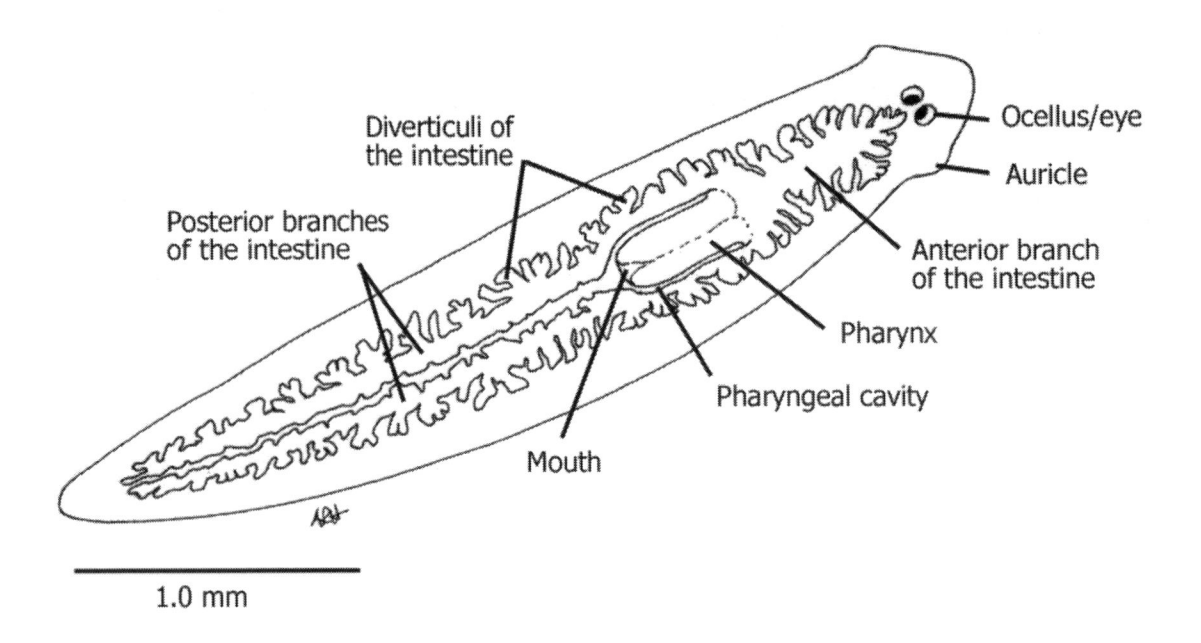

Figure 6.2. Digestive system of the triclad planarian *Dugesia*. (Image: ARH)

4) DRAW a cross-section slide through the pharynx of a planarian. Use **Fig. 6.3** to help you identify what you see.

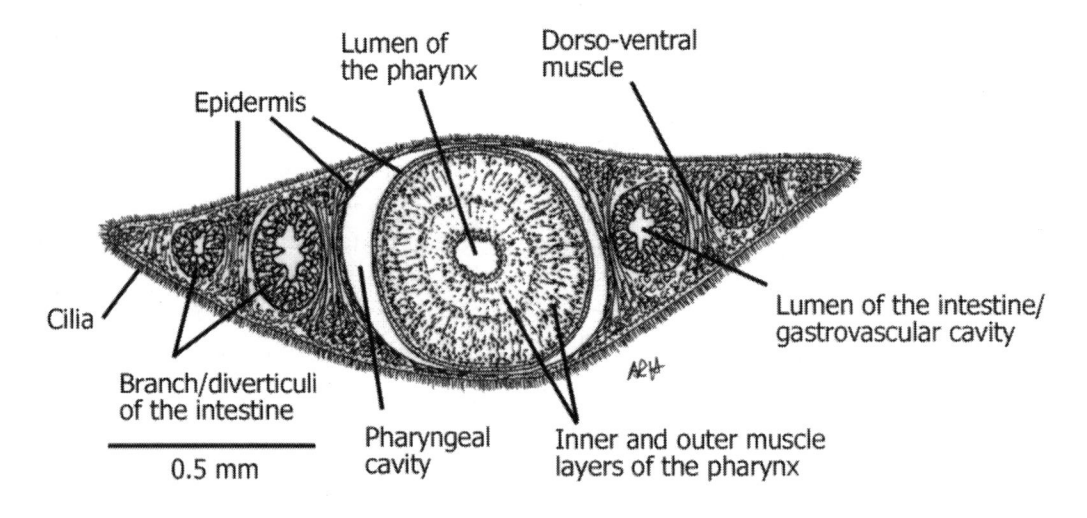

Figure 6.3. Cross-section through the pharynx of the triclad planarian *Dugesia*. (Image: ARH)

Infraclass Neodermata

All members of this group are parasitic and they go through a developmental process where they replace some or all of the original cellular epidermis with a syncytial outer covering called the neodermis or tegument. Neodermata includes Cohorts Monogenea, Digenea and Cestoda.

Cohort Monogenea - monogeneans

Monogenea are ectoparasitic flukes. The name Monogenea means "one birth" or "one origin." Monogeneans usually display a high degree of host species specificity. These flukes attach themselves to fins, gills and other thin tissues of their hosts with a unique posterior attachment structure called a haptor. Haptors have hooks or suckers or both. A monogenean then uses its mouth to penetrating their host's skin and feed on blood, cells and interstitial fluid. Monogeneans are small and usually have minimal effects on their hosts. When water temperatures increase however, monogenean reproductive rates increase and they can put significant stress on their hosts as their numbers increase and hosts become heavily infested.

Tasks – Cohort Monogenea

1) Sedate fish and look for monogeneans. Use stock MS222 anesthetic solution to anesthetize small fish and check for monogeneans (anesthetization procedure after Dr. Jason Hunt, Brigham Young University-Idaho, modified from a procedure developed by Dr. Edward E. Brandt, Shenandoah University).
    a. Make 100 ml of 0.4% MS222 stock solution
        - 400 mg MS-222 (Tricane)
        - 97.9 ml distilled water
        - 2.1 ml 1 M Tris-Cl (pH 9)
        - Add 5 ml MS-222 stock solution to 95 ml of clean tank water
    b. Net a fish and transfer it into 100 ml of the anesthetizing solution.
    c. When the fish becomes motionless remove it from the solution and do a dip rinse in distilled water to reduce risk of fish death.
    d. Place the sedated fish in a Petri dish or small glass bowl with just enough water to cover it. Be sure that the tail or a fin is near the center of the dish.
    e. Flare the tail or fin using toothpicks or insect pins and keep the tail spread during examination.
    f. Use a dissection scope to search for monogeneans. Begin with low magnification and focus on the surface of the tail or fins. If no parasites are observed on fins then examine gill tissue. Once monogeneans are located collect tissue parasites and make wet-mount slides for observation. Be sure you have generous amounts of plasticene clay on the corners of your coverslip so specimens are not smashed between the coverslip and slide.
2) DRAW a monogenean and describe what you see. Use **Fig. 6.4** to help you identify what you see.

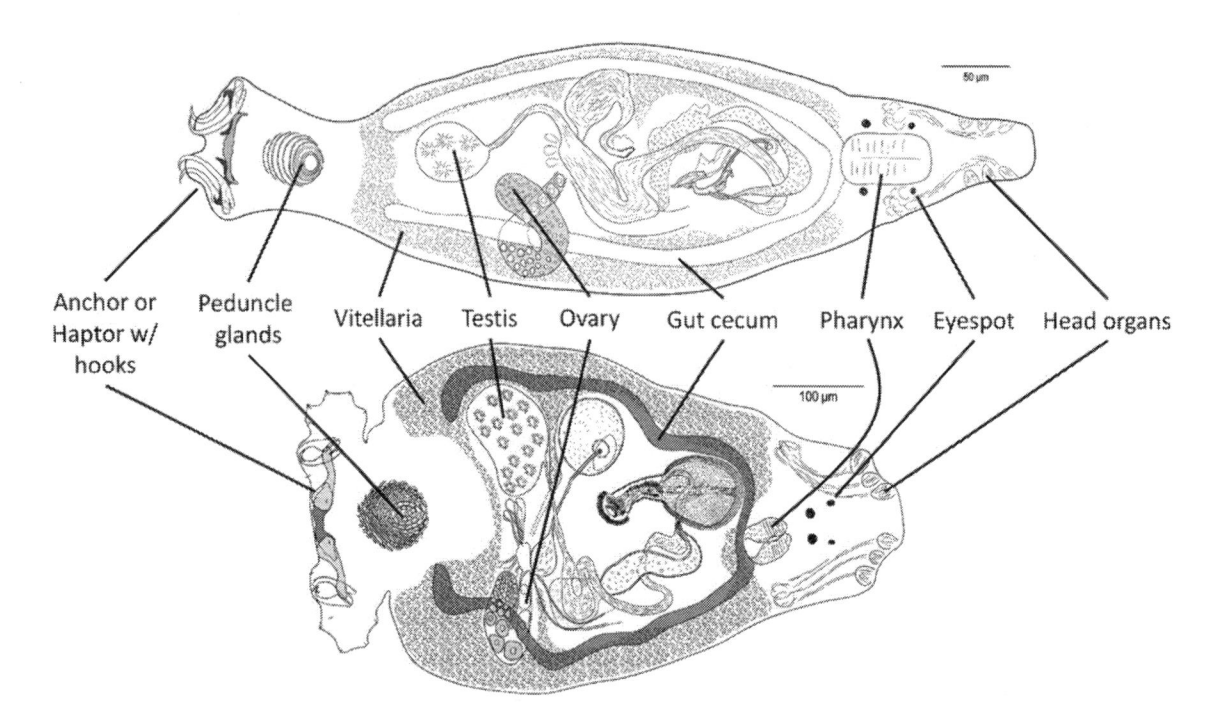

Figure 6.4. Representative Monogeneans. (Images: Jean-Lou Justine, under the Creative Commons Attribution-Share Alike 3.0 Unported License and the GNU Free Document License ver. 1.2)

Cohort Trematoda – digenean flukes

Digenean flukes have complex life histories, i.e., they require more than one host to complete their life cycles. The name digenea means "two births" or "two origins." Intermediate hosts house clonal life stages and sexually reproductive adults parasitize the definitive or final host.

This group is medically important; many kinds of digeneans parasitize humans. One digenean parasite of humans is the Chinese liver fluke *Clonorchis sinensis*. *Clonorchis* is common in areas of the world where the main crop is rice, human waste as used as fertilizer, farming is done largely via manual labor and people eat raw fish. The Chinese liver fluke life cycle is shown in **Fig. 6.5**.

Tasks – Cohort Digenea

1) DRAW the egg, redia, cercaria and metacercaria life stages of *Clonorchis* from prepared slides. Use **Fig. 6.6** to help you identify what you see. Also, locate each of these stages in the life cycle shown in **Fig. 6.5**.
2) DRAW an adult *Clonorchis* from a prepared slide. Use **Fig. 6.7** to help you identify what you see.

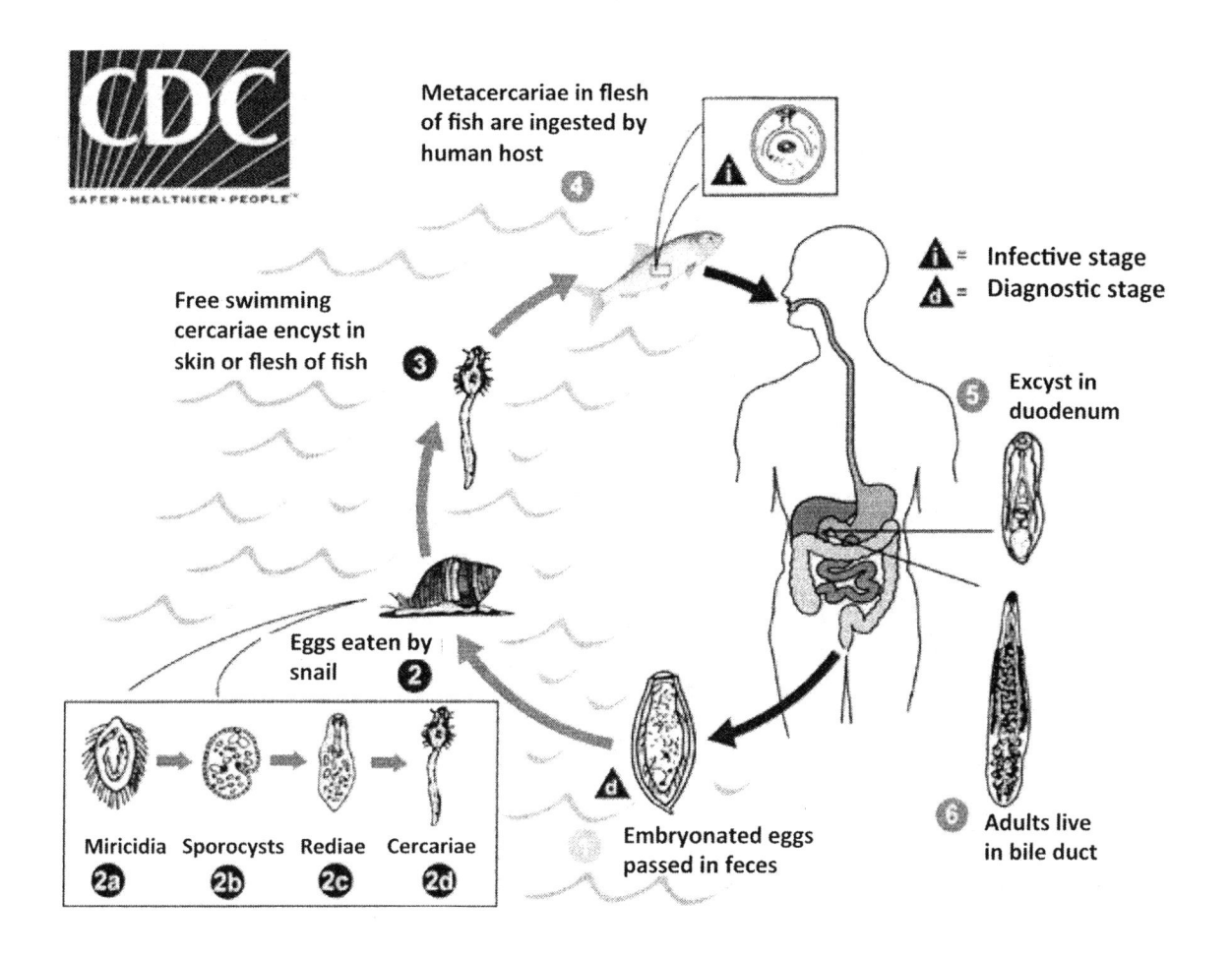

Figure 6.5. Life cycle of the Chinese liver fluke *Clonorchis sinensis*. The adult fluke lives in the human liver, gall bladder or bile duct (6). Adult flukes produce huge numbers of embryonated eggs that are released into the small intestine and exit the host body with the feces (1). Where human feces are used as fertilizer for rice, feces containing eggs are applied to rice paddies. Once in the water the feces break down and eggs float to the bottom where they are eaten by snails (2 a-d). Once inside a snail the embryo hatches as a miricidium stage larva. A miricidium develops into the sporocyst stage that produces many rediae internally via cloning. The sporocyst body wall eventually breaks and releases rediae into the snail body cavity. A single redia produces many cercariae also by internal cloning. Cercariae are also released into the snail body cavity. They burrow through the snail's body wall and swim until they come into contact with a suitable second intermediate host, typically a fish (3). A cercaria burrows into the skin or skeletal muscle of fish where it encysts and becomes a metacercaria (4). The life cycle is completed when the final host ingests raw or improperly cooked infected fish. The metacercaria excysts when the tissue surrounding it is digested away (5). The infective stage juvenile burrows through the gut wall and into the abdominal cavity of the final host. At this point the juvenile fluke crawls along the inner wall of the abdominal cavity in a random direction. It doesn't matter which way it goes because it will eventually come in contact with the liver where it burrows in and becomes an adult. (Image: CDC, http://www.cdc.gov/parasites/clonorchis/biology.html)

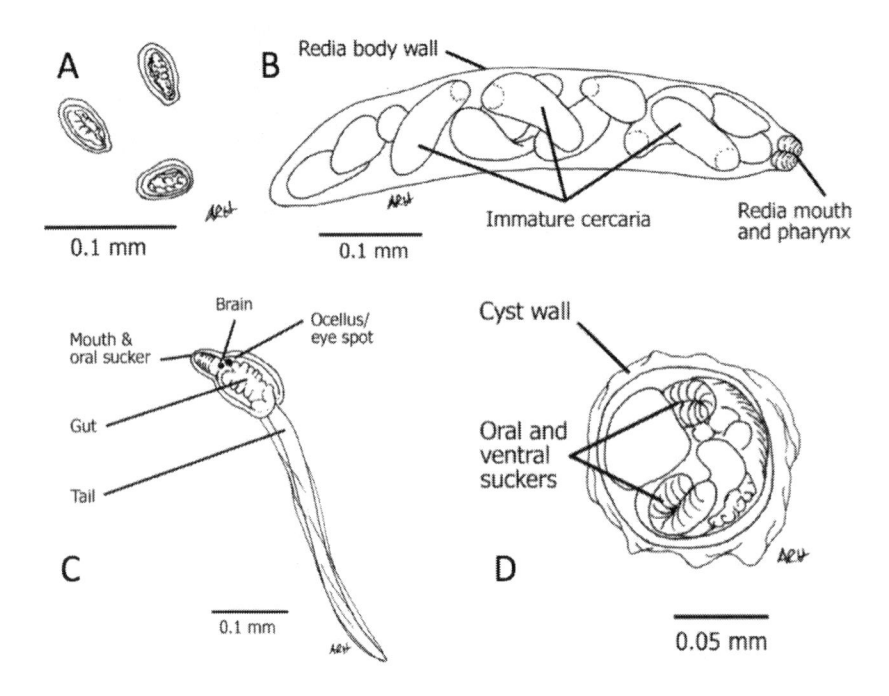

Figure 6.6. *Clonorchis sinensis* life stages: **A)** embryonated eggs, **B)** redia containing immature cercaria, **C)** cercaria, and **D)** metacercaria. (Image: ARH)

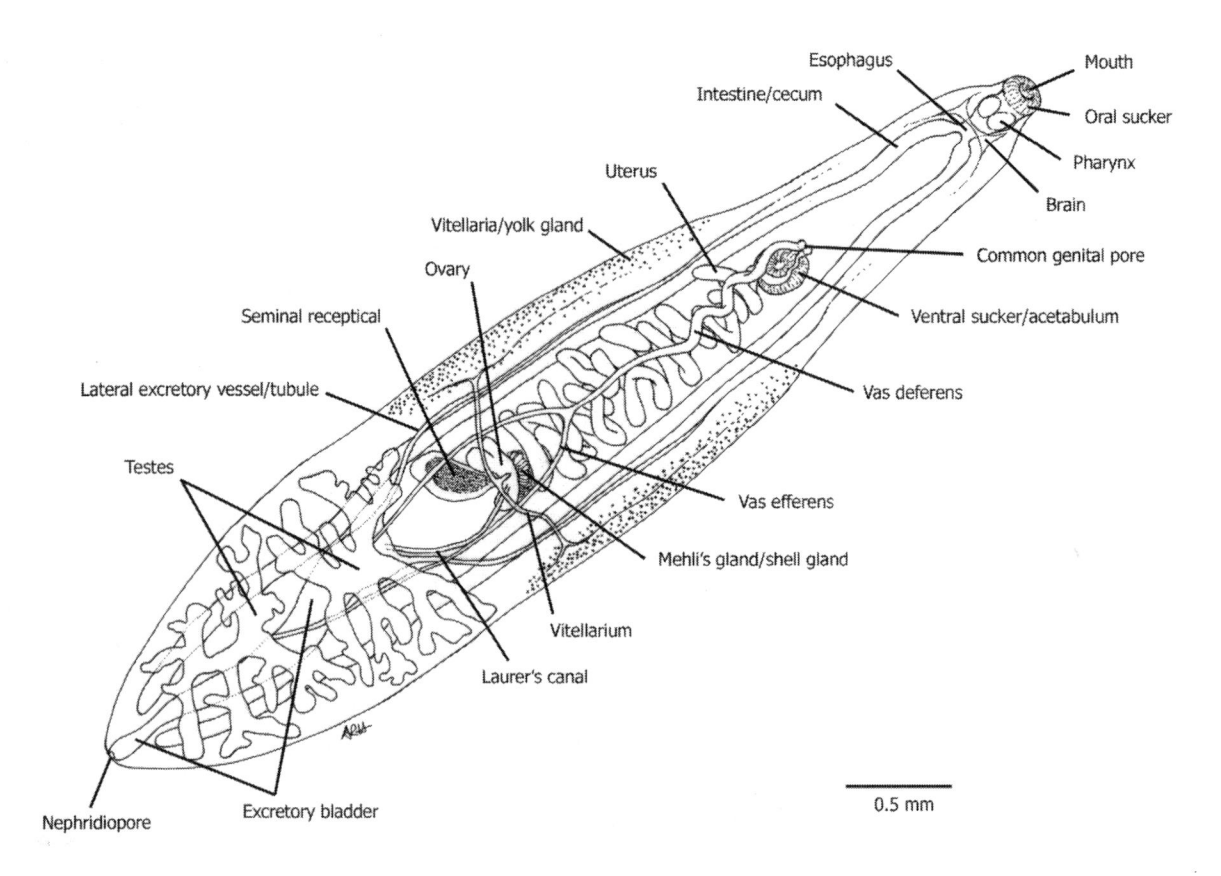

Figure 6.7. Adult *Clonorchis sinensis*. The uterus is packed with embryonated eggs, not shown here. (Image: ARH)

Cohort Cestoda - tapeworms

Tapeworms parasitize all species of vertebrates. The longest tapeworm lives in the intestines of whales and can be 30 m long. The longest tapeworms in humans are not much shorter at 25 m long, though most human tapeworms are much smaller.

Cestoda also have complex life histories. Their life histories are driven by links between herbivores and carnivores or opportunistic scavengers. The life cycle of the beef and swine tapeworms *Taenia saginata* and *T. solium*, species also found in humans is shown in **Fig. 6.8**. This cycle can also be completed by other species that eat beef, including dogs, wolves, rats, etc.

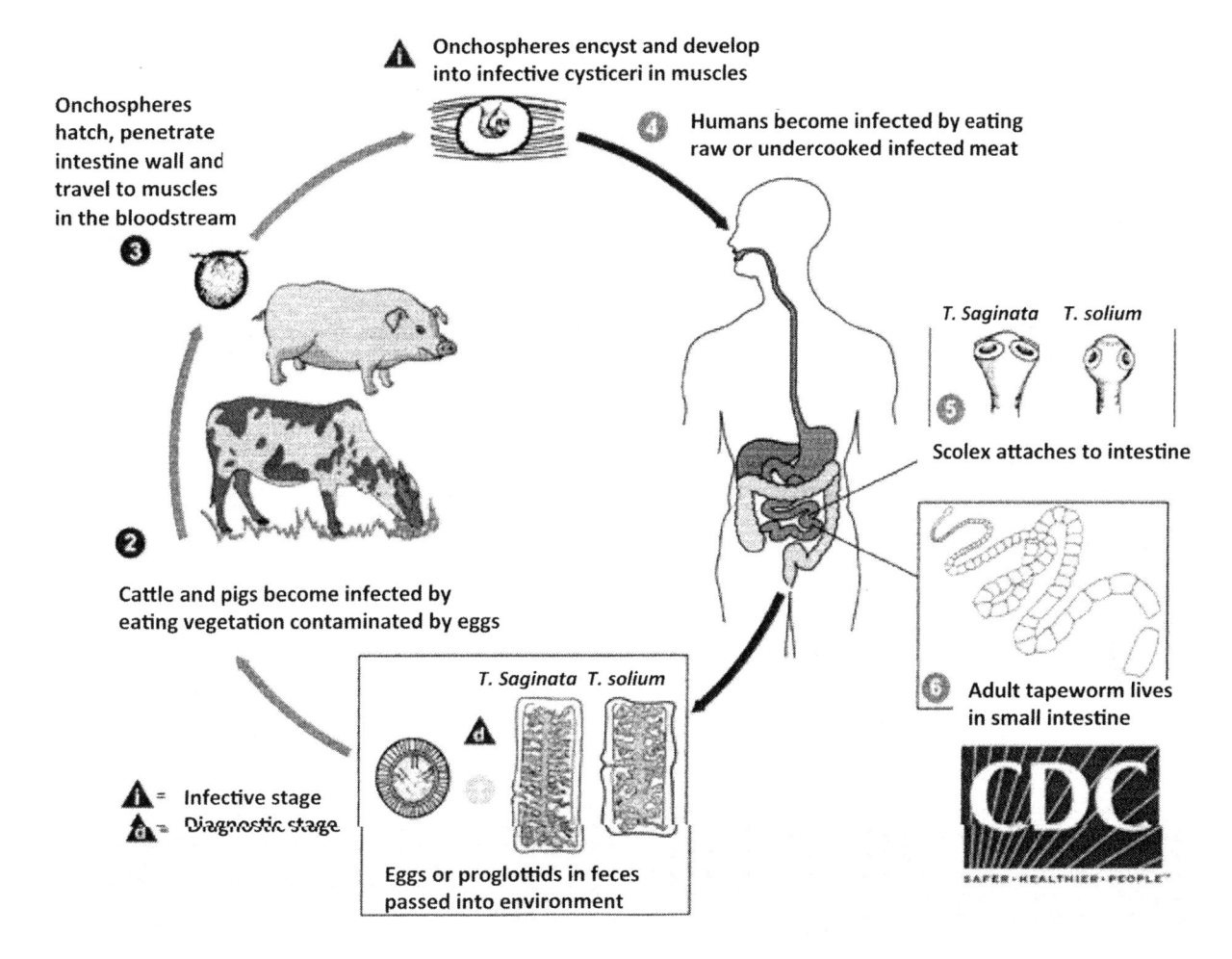

Figure 6.8. Life cycle of the tapeworms *Taenia saginata* and *T. solium*. Adult tapeworms live in the intestine of the definitive host and produce vast numbers of embryonated eggs (1). Eggs are released with the host feces. Some of these eggs are ingested by herbivores (2). Eggs hatch in the herbivore's gut and the infective oncosphere stage burrows into skeletal muscle where it encysts and develops into the cysticercus stage also called a bladder worm (3). When another animal digests infected tissue of an intermediate host the cysticercus stage excysts in the small intestine (4). The cysticercus everts its scolex and attaches to the wall of the host small intestine and metamorphoses into an adult (5-6). (Image: CDC, http://www.cdc.gov/parasites/taeniasis/biology.html)

Tasks – Cohort Cestoda

1) Observe prepared slides of the scolex, a mature proglottid and a gravid proglottid. DRAW these structures and use **Figs. 6.9 - 6.11** to help you identify what you see.

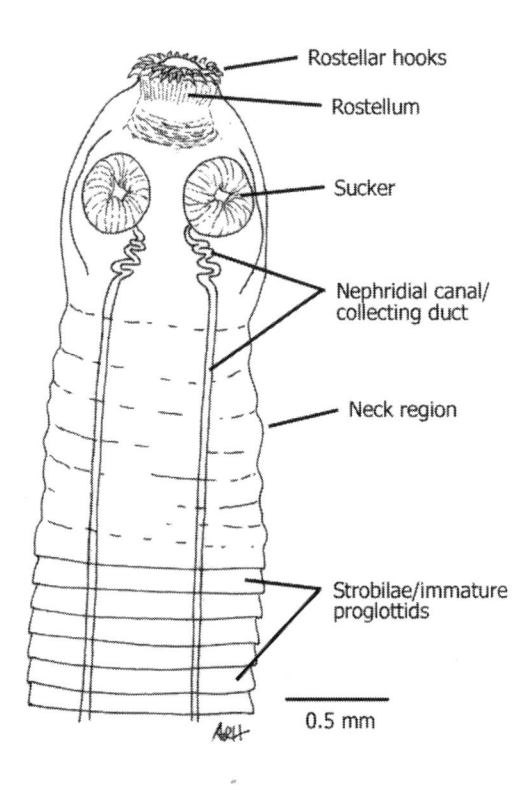

Figure 6.9. Scolex, neck and immature proglottids of *Taenia saginata*. (Image: ARH)

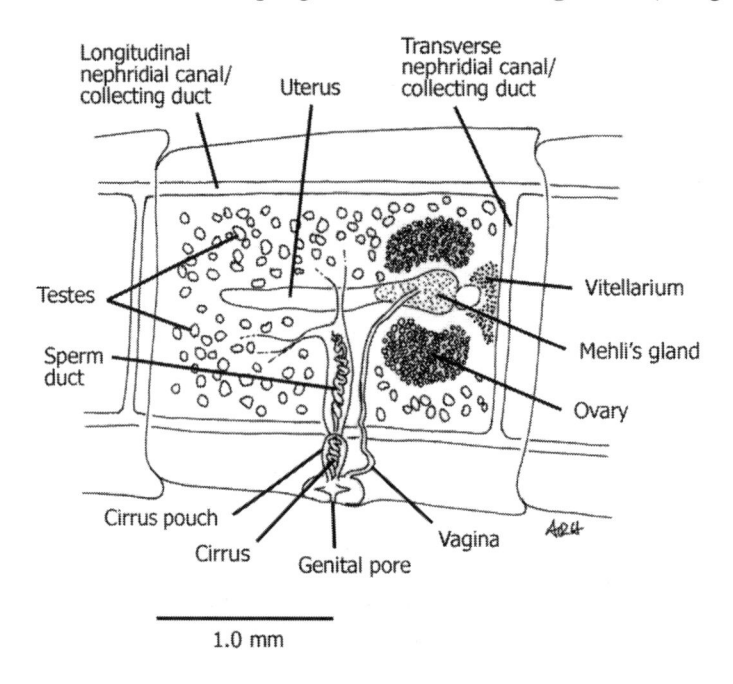

Figure 6.10. Mature proglottid of *Taenia saginata*. (Image: ARH)

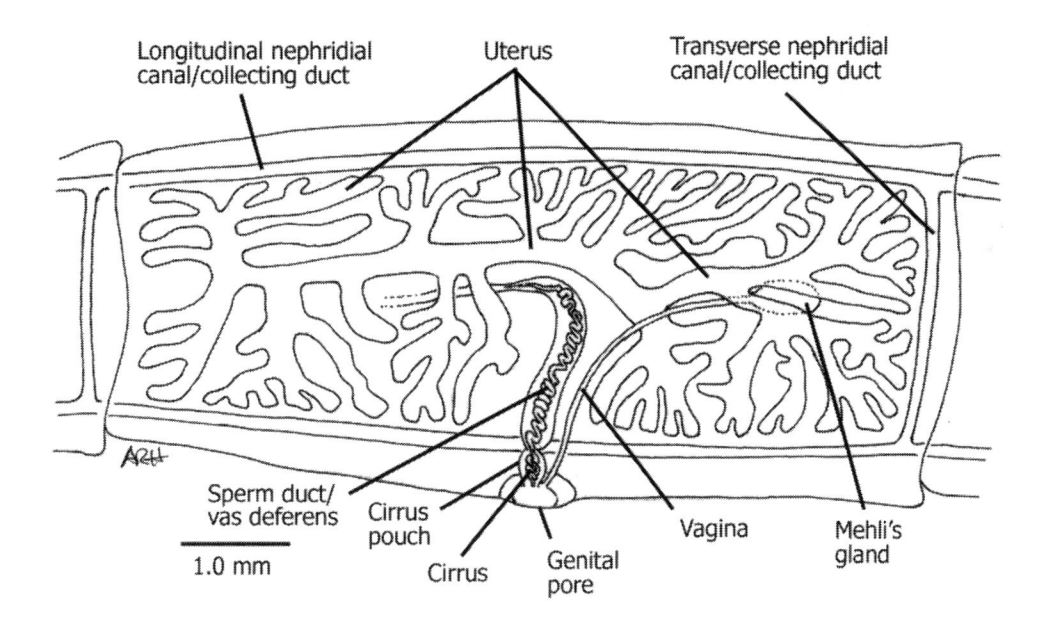

[Diagram of gravid proglottid with labels: Longitudinal nephridial canal/collecting duct, Uterus, Transverse nephridial canal/collecting duct, Sperm duct/vas deferens, Cirrus pouch, Cirrus, Genital pore, Vagina, Mehli's gland; scale bar 1.0 mm]

Figure 6.11. Gravid proglottid of *Taenia saginata*. Embryonated eggs are not shown in the uterus in this drawing, but the uterus in gravid proglottids will be completely packed with them. (Image: ARH)

Group Questions

1) List advantages that a bilateral body plan conveys that a radial body plan does not.
2) Develop a hypothesis that could explain how a parasitic lifestyle could have originated.
3) Why is it evolutionarily advantageous for most parasites to be monoecious?
4) Explain why even large flatworms from centimeters to meters long can do just fine without the complex circulatory systems seen in other large animals.
5) Why is the neodermis a beneficial adaptation for an endoparasite that resides in the digestive tract of its host?

Phylum Platyhelminthes – Glossary

Acoelomate body – triploblastic body plan where the only fluid-filled space in the body is the lumen of the gut

Auricle – lateral extensions of the head that bear chemosensory cells in some free-living planarians

Bilateral symmetry – body symmetry where mirror images can be obtained only by a longitudinal plane running along the midline of the body

Cephalization – concentration of nervous tissue and sensory structures in an anterior head

Cercariae – swimming life stage of digenean flukes, is produced within the redia stage and becomes the metacercaria in the intermediate host

Cirrus – male copulatory organ of tapeworms

Cirrus pouch – cavity that houses the tapeworm cirrus

Complex life history – a parasitic life history that requires two or more hosts

Cysticercus – tapeworm life stage that is encysted in intermediate host tissue and excysts and becomes an adult when its host tissue is digested

Definitive host / final host – the host that bears the adult life stage of a parasite

Diverticuli of the intestine – outpocketings or branches of the intestine that increase surface area and volume of the gut

Ectoderm – embryonic tissue that gives rise to the epidermis and nervous tissue

Ectoparasite – parasite that attaches to the external surface of its host

Embryonated egg – embryo produced by an adult parasite and is encased in an environmentally resistant outer covering or shell

Endoderm – embryonic tissue layer that gives rise to the digestive tract

Endoparasite – parasite that lives within the body of its host

Epidermis – tissue covering the outer surface of the body

Esophagus – tubular connection between the pharynx and the rest of the gut

Excretory bladder – organ in flukes where nitrogenous waste is stored prior to being released

Eyespot – see Ocellus

Genital opening / genital pore – opening in flukes and tapeworms where sperm are transferred and embryonated eggs are released

Gut cecum – blind extension of the gastrovascular cavity in flukes

Haptor – attachment organ of monogenean flukes that may have suckers, hooks or both

Laurer's canal – structure in flukes that functions like a vagina, is the tube through which sperm move to the seminal receptacle

Mehli's gland / shell gland – gland in flukes that produces the protective outer covering or shell for embryos, it also secretes mucus that provides lubrication that helps move encapsulated embryos through the uterus

Mesoderm – embryonic tissue layer that gives rise to muscles, mesenchyme and other tissues between the epidermis and gut

Metacercaria – life stage that excysts in the final host's gut and develops into an adult fluke

Miricidium – fluke life stage that emerges from the embryonated egg inside a snail and produces sporocysts

Nephridial canal / collecting duct – tubular excretory structure in adult tapeworms

Nephridiopore – opening through which nitrogenous waste is released from the body

Neodermis / tegument – syncytial tissue (large cell with many nuclei) that covers the external body surfaces of adult flukes and tapeworms

Ocellus / eye – cup-shaped photosensory structure that detects the presence, intensity and direction of light, it may be able to detect movement but it does not generate an image

Oncosphere – tapeworm life stage that develops from the excysted embryo, penetrates the lining of the gut and becomes the cysticercus stage

Oral sucker – muscular feeding and attachment organ in flukes, surrounds the mouth

Ovary – primary female reproductive organ, produces eggs

Pharyngeal cavity – space that houses the pharynx in free-living flatworms

Pharynx – muscular tube used to pull material into the gut

Proglottid – Mature – compartmentalized body region of tapeworms that carries out copulation produces embryonated eggs; Gravid - compartmentalized body region of tapeworms that has an extensive uterus that is packed with embryonated eggs

Protonephridia – excretory organ containing one to several flagellae that pull interstitial fluid into a tubule where the fluid is modified via absorption and secretion to create final urine

Redia – fluke life stage that develops within the sporocyst and gives rise to cercariae

Rostellar hooks – hooks of the scolex help attach a tapeworm to the intestinal wall of its host

Rostellum – attachment organ of a tapeworm, may have suckers, hooks or both

Seminal receptacle – organ that stores sperm after copulation

Simple life history – a parasitic life history that includes only one host

Sperm duct / vas deferens – duct in tapeworms that collects sperm from the testis and stores it until copulation

Sporocyst – fluke life stage that develops from the miricidium and produces rediae internally via cloning

Testis – primary male reproductive organ, produces sperm

Triploblastic – body plan that produces three embryonic tissue layers: ectoderm, endoderm and mesoderm

Uterus – organ that houses embryonated eggs until they are released

Vagina – opening and tube through which sperm are received during copulation

Vas deferens – tubule in flukes that is formed when the vas efferens fuse and carries sperm to the common genital pore

Vas efferens – duct in flukes leaving the testis that carries sperm to the vas deferens

Ventral sucker / acetabulum – organ of attachment flukes, is located on the ventral surface of the body and may have accessory spines

Vitellarium – tissue/organ that produces yolk

Chapter 7: Phylum Mollusca

Estimates vary but there are at least 80,000 described species of molluscs. Phylum Mollusca includes a great diversity of specialized forms that are all derived from a common ancestral mollusc body plan. There are tiny slow creepers and large fast swimmers, and molluscs make a living in every imaginable way: predators, herbivores, scavengers, suspension feeders, parasites and a few even have symbiotic relationships with photosynthetic zooxanthellae.

Most molluscs are marine but some gastropods and bivalves live in freshwater and a few species of snails and slugs are terrestrial. Characteristics of molluscs are listed in **Table 7.1**.

Table 7.1. Characteristics of Phylum Mollusca (after Brusca, 2016).

- Triploblastic, bilateral, unsegmented coelomate body
- Visceral mass
- Mantle that often includes a shell gland that produces a calcified shell, shell plates or spicules
- Mantle cavity
- Muscular foot
- Radula and odontophore complex (in many groups)
- Complete gut
- Metanephridia
- Trochophore larva (in many groups)

Phylum Mollusca "shellfish" - Taxonomy

Phylum Mollusca includes eight classes though some taxonomists assign the first two classes in this list to a single taxon called Class Aplacophora "no shell bearing".

Follow this link to watch an introduction to the anatomy and lifestyles of molluscs from *The Shape of Life* video series: http://shapeoflife.org/video/molluscs-survival-game.

Class Caudofoveata "foot pit" (not covered in this exercise) – aplacophorans also called spicule worms. These strictly marine worm-like molluscs secrete a chitinous cuticle and calcareous spicules but no shell. They lack a foot but have a pair of ctenidia in a small posterior mantle cavity and a radula is always present.

Class Solenogastres "pipe belly" (not covered in this exercise) – also aplacophorans and also called spicule worms. These strictly marine molluscs also produce a chitinous cuticle and calcareous spicules as well as a foot groove but they lack ctenidia and a shell, the radula is lacking in some species.

Class Monoplacophora "one shell bearing" (not covered in this exercise) – monoplacophorans. These small snail-like marine molluscs produce a cap-shaped shell and have a circular foot with eight pairs of foot retractor muscles. There are three to six pairs of ctenidia located along the edges of the foot. Most monoplacophorans are found in the deep sea and until the early 1950s were known only from fossils.

Class Polyplacophora "many shell bearing" (covered in this exercise) – chitons. These molluscs are strictly marine and have a dorso-ventrally flattened body with most of the dorsal surface covered by eight overlapping shell plates. These plates are embedded partially to completely in a tough fleshy girdle. Shell plates have unique canals in them called aesthetes that sometimes function as light sensory organs. Chitons have a broad muscular foot with long groove-like mantle cavities running along the length of the body between the foot and the girdle and contain up to 80 pairs of ctenidia. The head lacks tentacles and has a well-developed radula. Most chitons make a living by using the radula to scrape algae or bacterial film off of rock surfaces or to feed directly on algae. Chitons can be locally abundant in rocky intertidal habitats. This short video shows the feeding behavior of *Placiphorella velata,* an usual predaceous chiton: https://www.youtube.com/watch?v=4bvTffG90rE

Class Gastropoda "stomach foot" (covered in this exercise) – snails, slugs and limpets. This is the most diverse group of molluscs with about 70,000 described species. Gastropods live in marine, freshwater and terrestrial environments. The diversity of their lifestyles is equally diverse and includes predators, herbivores, scavengers and even parasites. All members of this class carry out a developmental process called torsion in the larval stage. During torsion the visceral mass rotates $90\text{-}180^0$ with respect to the head/foot, but in slugs torsion is followed by detorsion. In snails the effect of torsion is to rotate the mantle cavity to an anterior position where the head can be quickly withdrawn into the shell. Gastropods have a strong creeping foot, one or two pairs of cephalic tentacles and a radula and odontophore complex. Aquatic forms have ctenidia in the mantle cavity while terrestrial forms have a vascularized lung-like mantle cavity. This short video from the BBC shows how a predaceous cone shell captures fish: https://www.youtube.com/watch?v=FYh2zeAsRXY

Class Bivalvia "two shells" (covered in this exercise) – clams, mussels and relations. The bivalves include marine and freshwater species and are some of the most highly derived molluscs. Freshwater bivalves are particularly important as indicators of environmental quality and many species are threatened or endangered. Bivalves, commonly referred to as clams, have two calcareous shells connected to each other by a ligament made of tough but stiff and flexible protein. Large adductor muscles close the shells and the ligament opposes these muscles so shells gape slightly when the adductor muscles relax or the clam dies. Clams lack a head and have a greatly enlarged mantle cavity that houses a large single pair of ctenidia. The foot is modified into a spade-like structure that can be used for burrowing in soft sediment. The head is highly reduced; it's no longer evident except for the presence of the mouth. Most bivalves make a living by suspension feeding by pulling water into the mantle cavity and using the ctenidia to capture suspended particles.

Class Scaphopoda "hollow foot" (not covered in this exercise) – tusk shells. These animals secrete a tapering tubular shell with an opening at both ends and may be up to 15 cm long. These animals live at the sediment-water interface. The smaller excurrent end of the shell extends above the sediment layer and expels water and waste while water is

pulled into the larger buried end. Movement of water through the tubular mantle pulls oxygenated water downward into the sediment, oxygenating the sediment and facilitating the growth of bacteria and protozoans that scaphopods use for food. There is a muscular foot used for burrowing and many ciliated threadlike captacula used to collect food particles. This short video shows a tusk shell using its foot to burrow through a clear matrix so the pattern of burrowing is visible:
https://www.youtube.com/watch?v=ENsZ5SLL3bs

Class Cephalopoda "head foot" (covered in this exercise) – octopus, squid and relations. This group includes the most intelligent invertebrates on the planet. They exhibit memory, learning and problem-solving capabilities. They have a relatively large brain-to-body ratio and have evolved image-forming lens-bearing eyes. In many species the mantle is modified into a muscular sac that is contracted forcefully producing jet propulsion, e.g., octopus and squid. The foot is modified into prehensile arms and tentacles with suckers or hooks. Cephalopods use a powerful jaw-like beak and toxins to kill prey and a radula to grind up food before it is swallowed. Most cephalopods have chromatophores in their skin that they can use for camouflage or communication. This video from Public Radio International's Science Friday program shows the camouflage capabilities of octopus: https://www.youtube.com/watch?v=eS-USrwuUfA

Class Polyplacophora - chitons

There are over 900 species of Polyplacophorans, commonly called chitons. Chitons range in size from less than 1 cm to over 30 cm in length. Eight overlapping shell plates and a leathery girdle protect the dorsal surface of chitons. Chitons also have a large, broad foot and a reduced head.

Tasks – Class Polyplacophora

1) Examine the external anatomy of your specimen. DRAW the dorsal and ventral surfaces of your specimen (**Fig. 7.1**).

Figure 7.1. Ventral view of the giant gumboot chiton *Cryptochiton stelleri*. (Image: ARH)

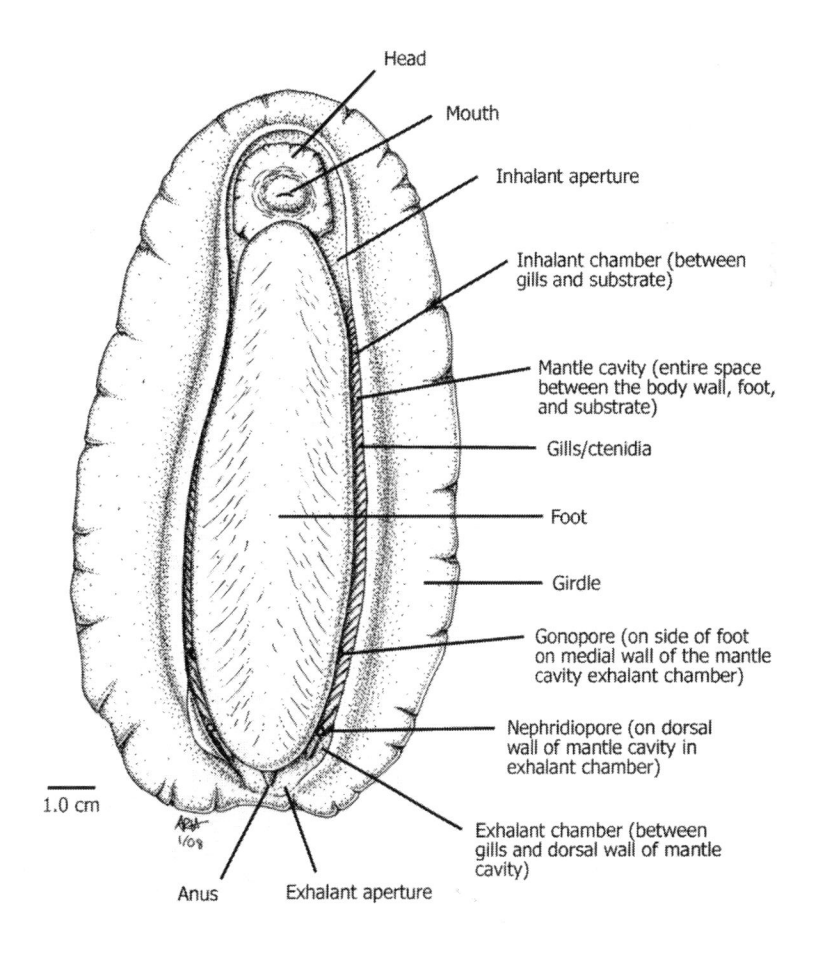

2) Work in pairs to complete the following dissection.
   a) Place your chiton ventral surface up. Use a scalpel or razor blade to make a shallow cut along the mid-line of the foot. Do not damage any underlying structures. Note the muscular texture of the foot. Carefully remove the entire foot, the gills and mantle cavity, thus exposing the body cavity. Rinse and then immerse your specimen. DRAW what you see in the body cavity. Refer to **Fig. 7.2** to help you identify what you see.

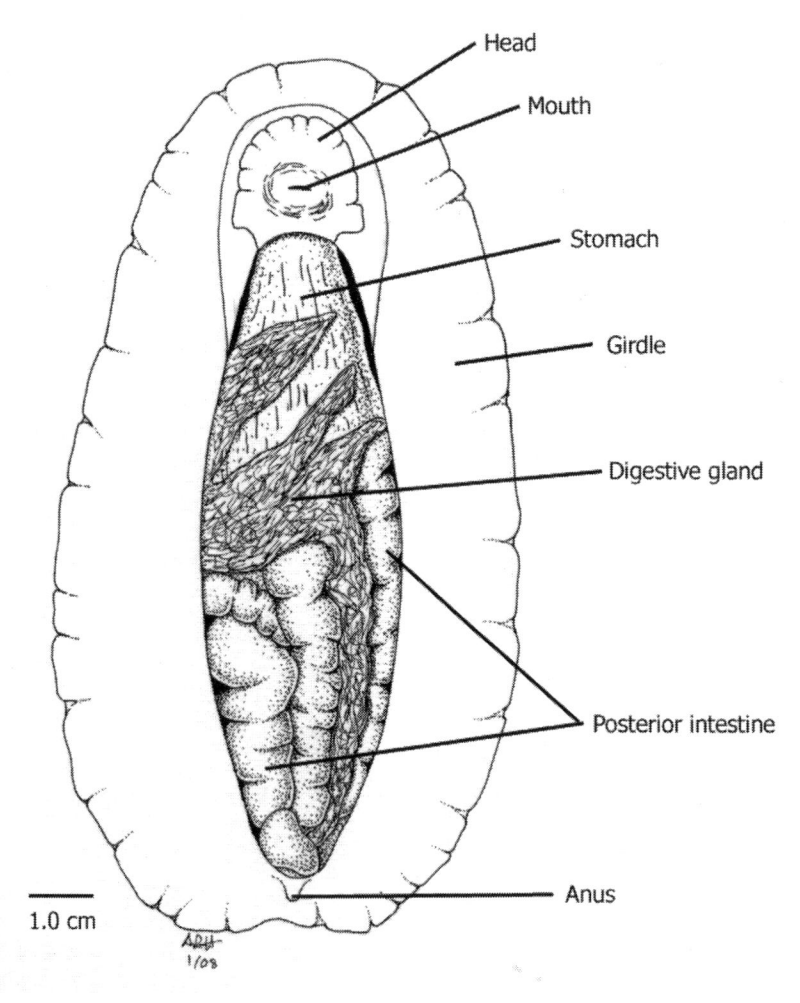

Figure 7.2. Ventral view of *Cryptochiton* with the foot and mantle cavities removed. (Image: ARH)

   b) Remove the digestive gland. It is the material between the loops of the intestine. Take your time removing this organ because walls of the intestine are easily damaged. DRAW the digestive tract after the digestive gland has been removed. Refer to **Fig. 7.3** to help you identify what you see.
   c) Carefully uncoil the intestine to see how long it is and observe how much of the body cavity is devoted to the digestive system. Next, cut around the edges of the anterior-most portion of the stomach and remove the entire digestive tract. If you are careful you should see the radular sac extending posteriorly from a mass of tissue at the anterior end of the body cavity. You should also be able to see the heart dorsally at the posterior end of the

body cavity. Observe the gonad that lies along the dorsal wall of the body cavity. Nephridia lie along the lateral walls of the body cavity. You should also see a pair of branching salivary glands lying along the body wall dorsal and lateral to the radular sac. DRAW what you see. Refer to **Fig. 7.4** to help you identify what you see.

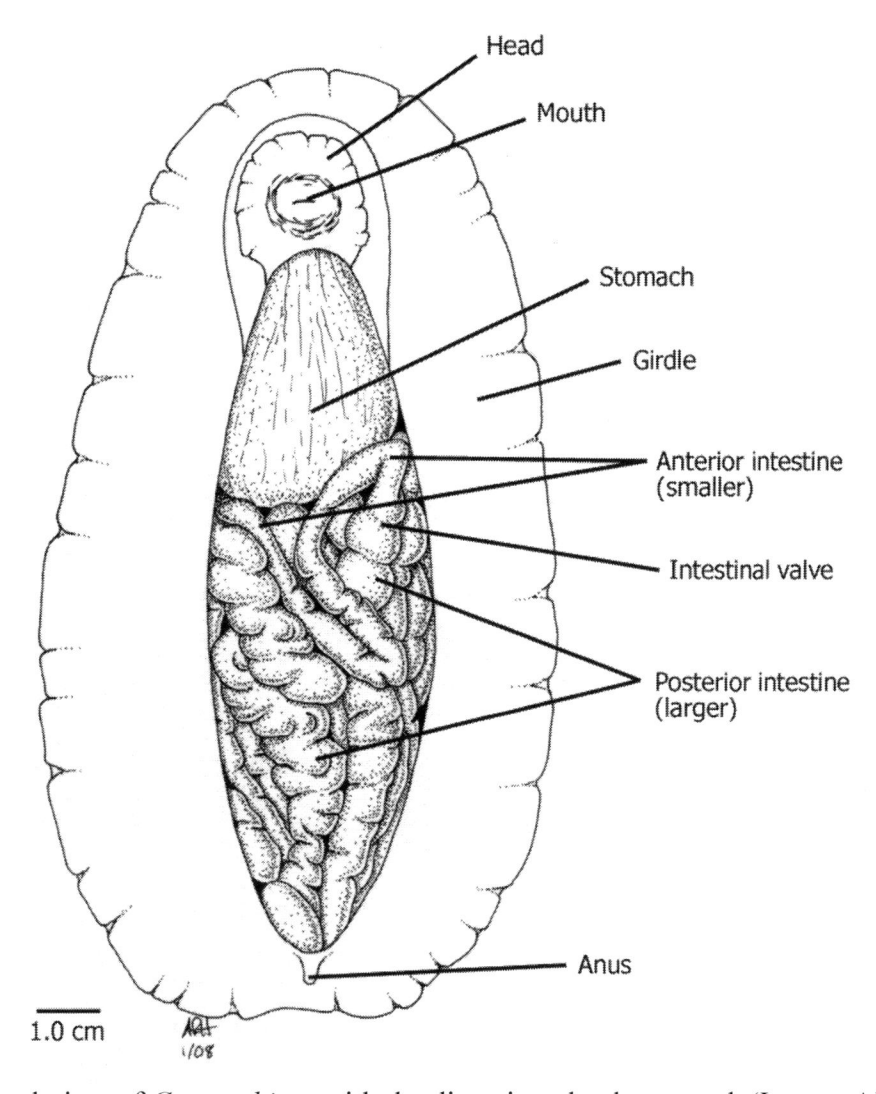

Figure 7.3. Ventral view of *Cryptochiton* with the digestive gland removed. (Image: ARH)

d) Remove the gonad and observe the inner surface of shell plates, dorsal aorta, muscles and other structures of the dorsal body wall.
e) Dissect the head. Use a scalpel to remove the head and associated tissues entirely. Make a cut through the midline of the head so you can lay it open and observe mirror images of the structures of the head. You should be able to see the buccal cavity, radula, radular sac, and odontophore. DRAW what you see (refer to **Fig. 7.5**).
f) Remove the radula and examine it with a magnifying class or dissection scope. DRAW what you see.

86

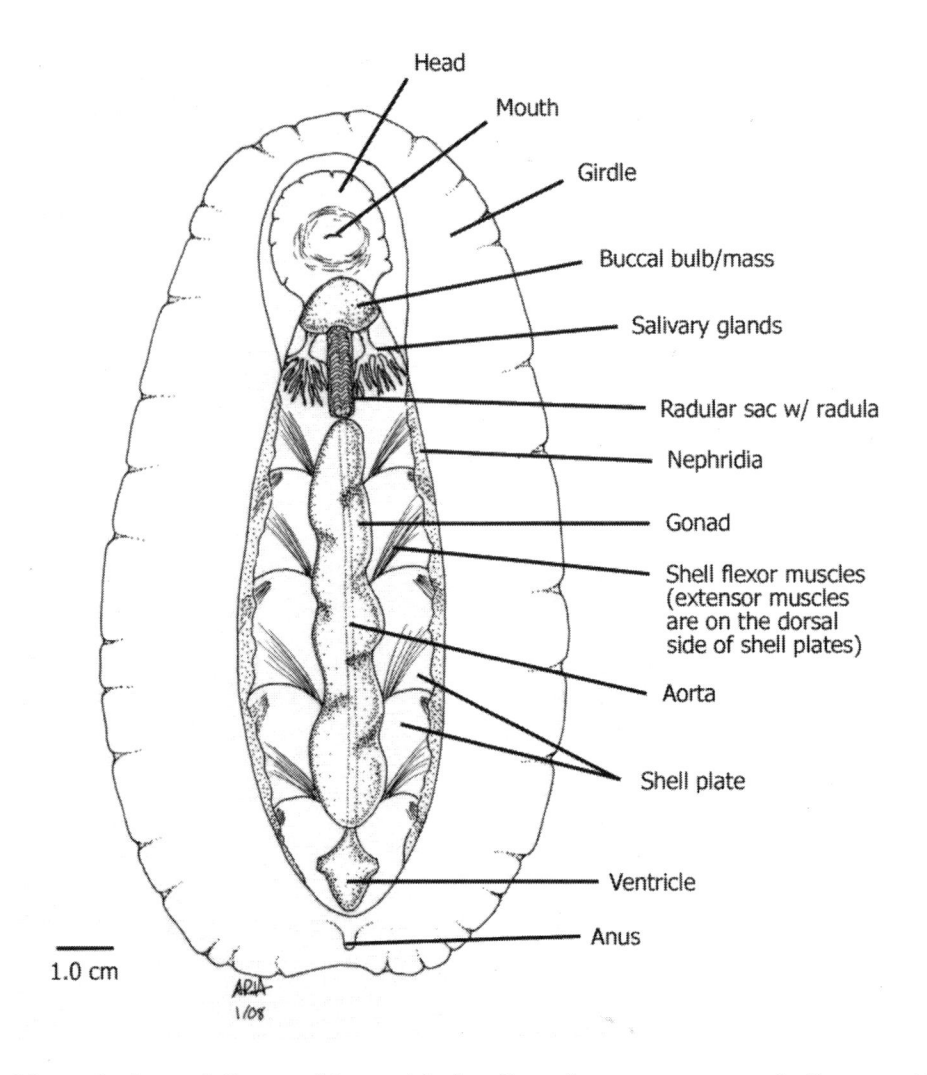

Figure 7.4. Ventral view of *Cryptochiton* with the digestive tract removed. (Image: ARH)

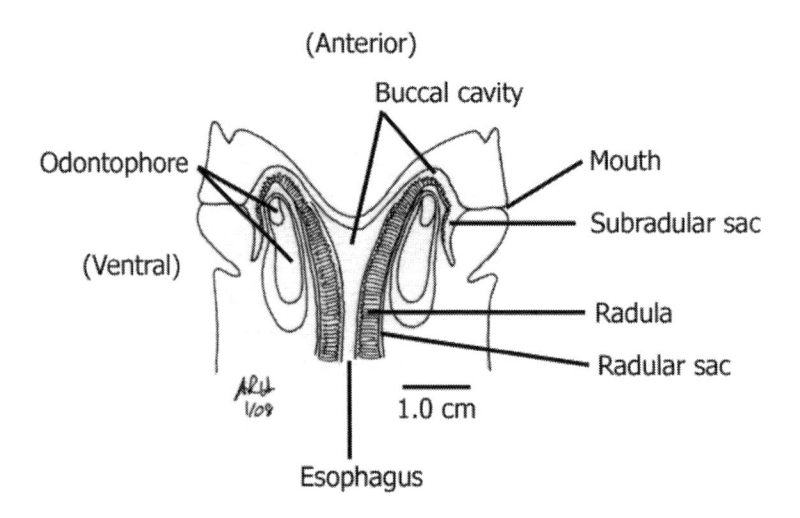

Figure 7.5. Head, buccal cavity and radular structures of *Cryptochiton*. (Image: ARH)

Class Gastropoda – snails and slugs

In this exercise you will study the anatomy of the terrestrial snail *Helix*. *Helix* is an economically important genus; some species in this group are used to make the French delicacy escargot.

Tasks – Class Gastropoda

1) Observe the external anatomy of *Helix*. Note the thickness of the shell and structures of the head, foot and collar. Compare the shell of *Helix* to those of marine snails. DRAW what you see (refer to **Fig. 7.6**).

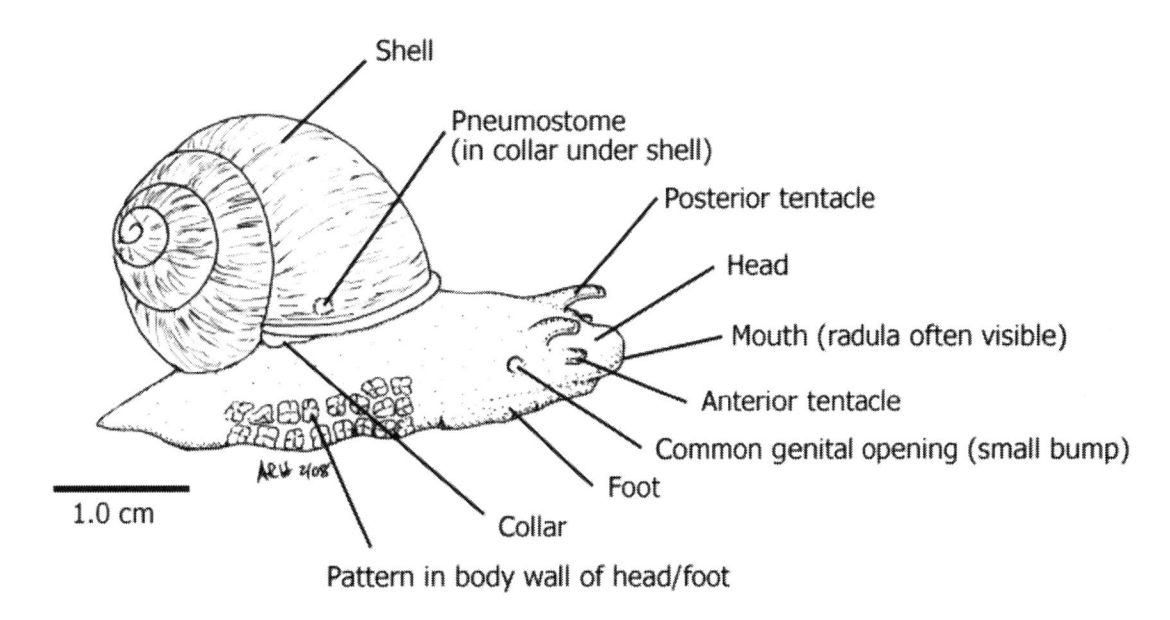

Figure 7.6. External anatomy of *Helix*. (Image: ARH)

2) Remove the shell. Use forceps to carefully crack and peel the shell away from the soft tissues of the body. Do not damage the soft tissues. Take your time because if you don't your snail could turn quickly into a mangled mess. Once the shell is removed place your specimen in a wax-bottom tray and fill it with enough water to cover your specimen. Use insect pins to pin the posterior portion of the foot and the anterior margin foot to the tray. Observe the structures that were exposed when the shell was removed. DRAW what you see and refer to **Fig. 7.7** to help you identify visible structures.

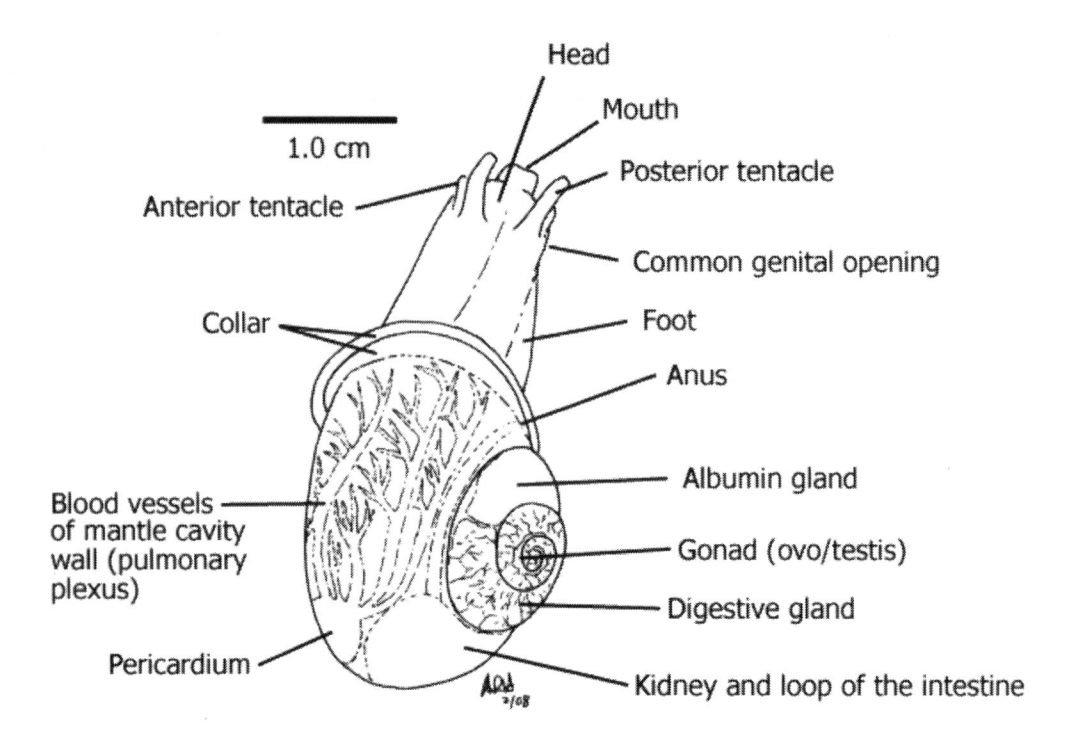

Figure 7.7. Dorsal view of *Helix* with the shell removed. (Image: ARH)

3) Open the mantle cavity. Insert the tip of a pair of scissors into the pneumostome and cut the tissue connecting the collar to the body wall of the snail. Continue your cut until you have separated the collar and the left margin of the mantle from the body wall. Fold the dorsal wall of the mantle cavity over so you can see the structures of the inner surface of the dorsal wall of the mantle cavity as well as structures visible through the floor of the mantle cavity. A network of network of blood vessels should be visible in the wall of the mantle cavity. This is the pulmonary plexus, the site of gas exchange. DRAW what you see (refer to **Fig. 7.8**).

Figure 7.8. Dorsal view of *Helix* with the dorsal wall of the mantle cavity folded over to the right. (Image: ARH)

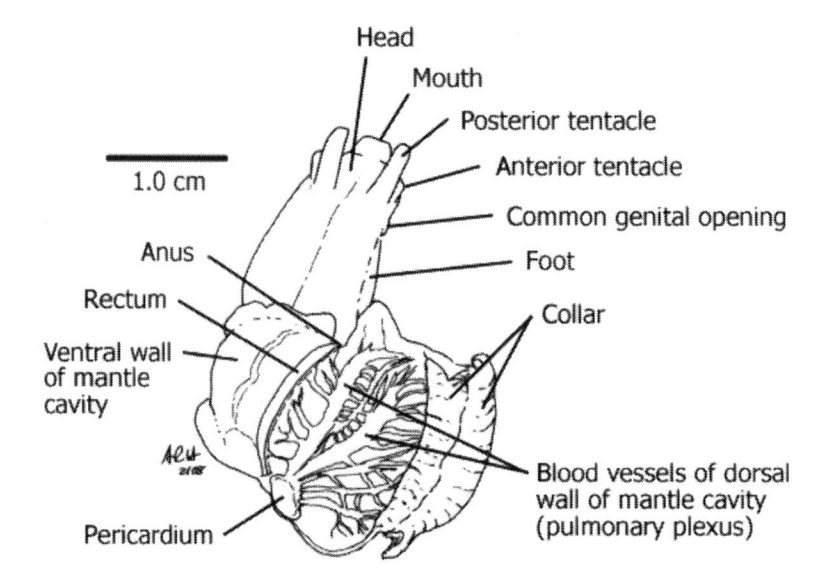

4) Open the body cavity. Use sharp-tipped scissors and fine-tip forceps to make a longitudinal cut through the dorsal wall of the head/foot starting at the mantle cavity and ending at the anterior end of the head. Pull the tissues of the body wall carefully away from each other and use insect pins to anchor the body wall to the floor of the dissection tray. Examine exposed structures of the body cavity before teasing anything apart. These structures comprise mainly the digestive system on the left side of the body and the reproductive system on the right side of the body. There are also many muscles, mostly along the floor of the body cavity, but these are not identified in this exercise. DRAW what you see. Refer to **Fig. 7.9** to help you identify visible structures.

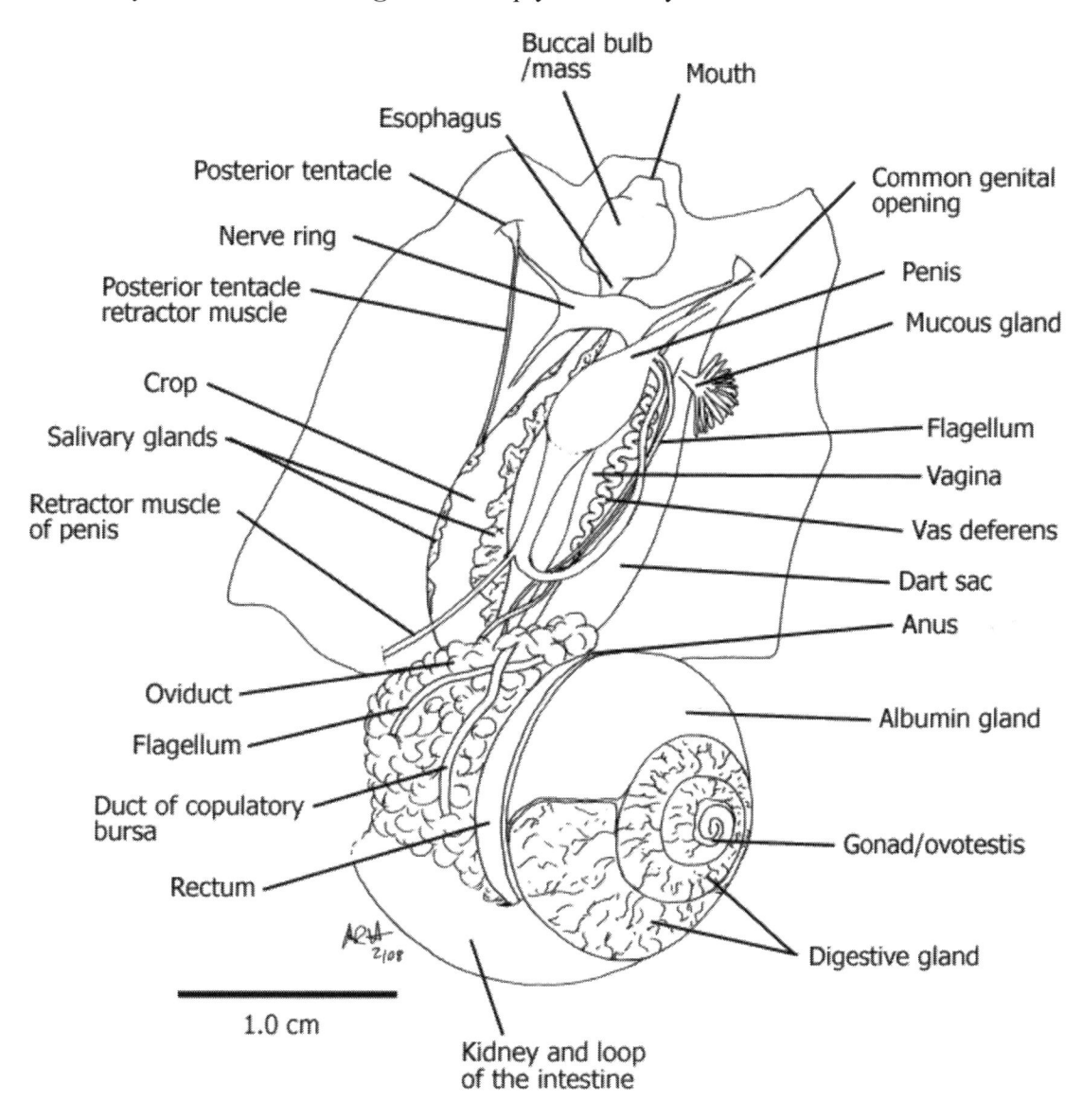

Figure 7.9. Internal anatomy of *Helix*, dorsal view. (Image: ARH)

5) Remove the oviduct and observe the anatomy of the posterior portion of the body cavity and visceral mass. Be careful not to tear or rip anything. DRAW what you see and refer to **Fig. 7.10** to help you identify visible structures.

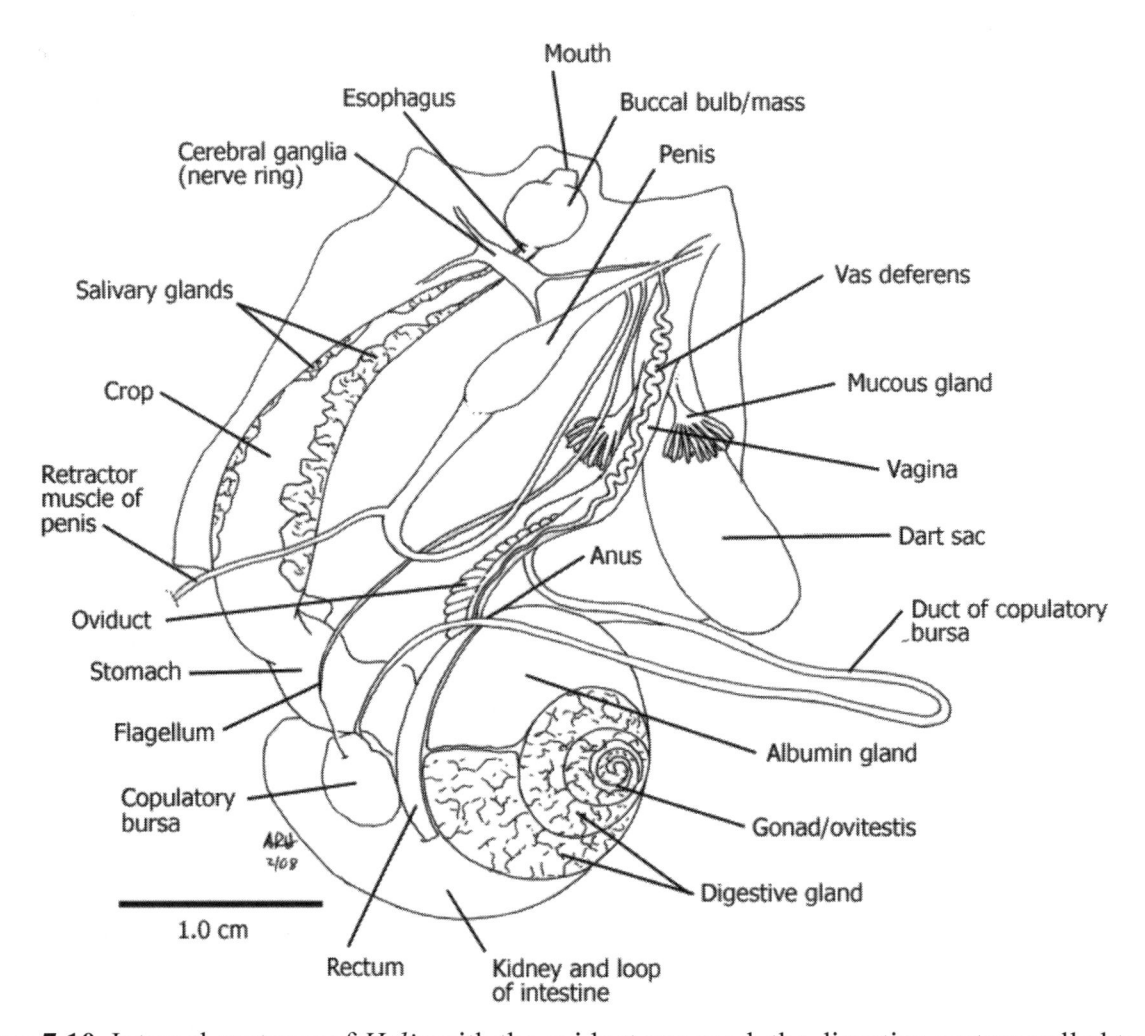

Figure 7.10. Internal anatomy of *Helix* with the oviduct removed, the digestive system pulled to the left and the reproductive system pulled to the right. Muscles and other structures of the floor of the body cavity are not shown. (Image: ARH)

Class Cephalopoda – octopus, squid and relations

Cephalopods are intelligent sight predators. As mentioned earlier they exhibit learning, memory and problem solving both by trial and error and by observing other individuals solve problems. This video footage from a field study in False Bay, South Africa, shows an octopus in the wild figuring out how to steal a bait-box intended to attract fishes: https://www.youtube.com/watch?v=T5tPAYx-Bmo

Tasks – Class Cephalopoda

1) Examine the external anatomy of a squid. Identify the morphological and functional polarity of its body. WRITE in your lab notebook the difference between the anatomical

and functional polarity of the squid body (anterior-posterior, dorsal-ventral, and medial-lateral). DRAW the external anatomy of your squid and indicated clearly both the anatomical and functional polarity of your specimen. All instructions and figures in this exercise refer to squid <u>functional polarity</u>. DRAW what you see and use the drawings in **Fig. 7.11** to help you identify structures of the external anatomy of a squid.

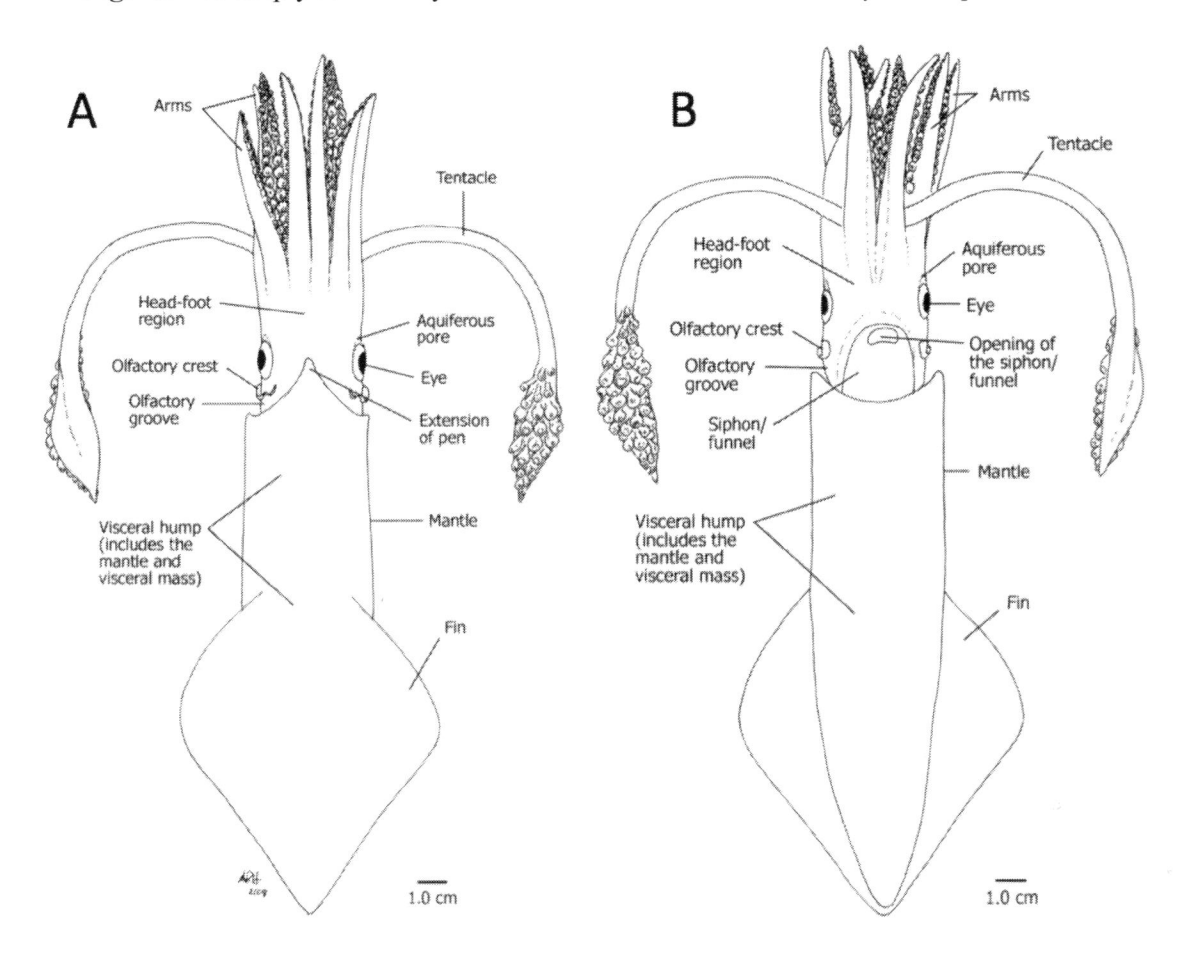

Figure 7.11. External anatomy of squid: **A)** Dorsal view, **B)** Ventral view. (Images: ARH)

2) Anatomy of the visceral mass. Place the squid on its back. Make a longitudinal incision through the ventral wall of the mantle. Make your incision off-center of the ventral midline. Start the incision at the anterior lip of the mantle and cut all the way to the posterior end of the mantle. Open the mantle cavity carefully. About 1/2 to 2/3 of the distance back from the anterior edge of the mantle you should see a blood vessel connecting the visceral mass to the ventral mantle wall. This is the medial mantle artery. You should also be able to see the medial septum that starts at the medial mantle artery and continues to the posterior end of the mantle cavity. The medial mantle artery and the medial septum are delicate structures and will disconnect easily from their points of attachment when you open the mantle cavity if you are not careful. After identifying the medial mantle artery and medial septum open the mantle cavity and use heavy dissection pins to anchor the walls of the mantle to the bottom of your dissection tray or cut the walls of the mantle away to keep them from flipping closed. DRAW the anatomy of the mantle cavity and refer to **Fig. 7.12** to help you identify what you see.

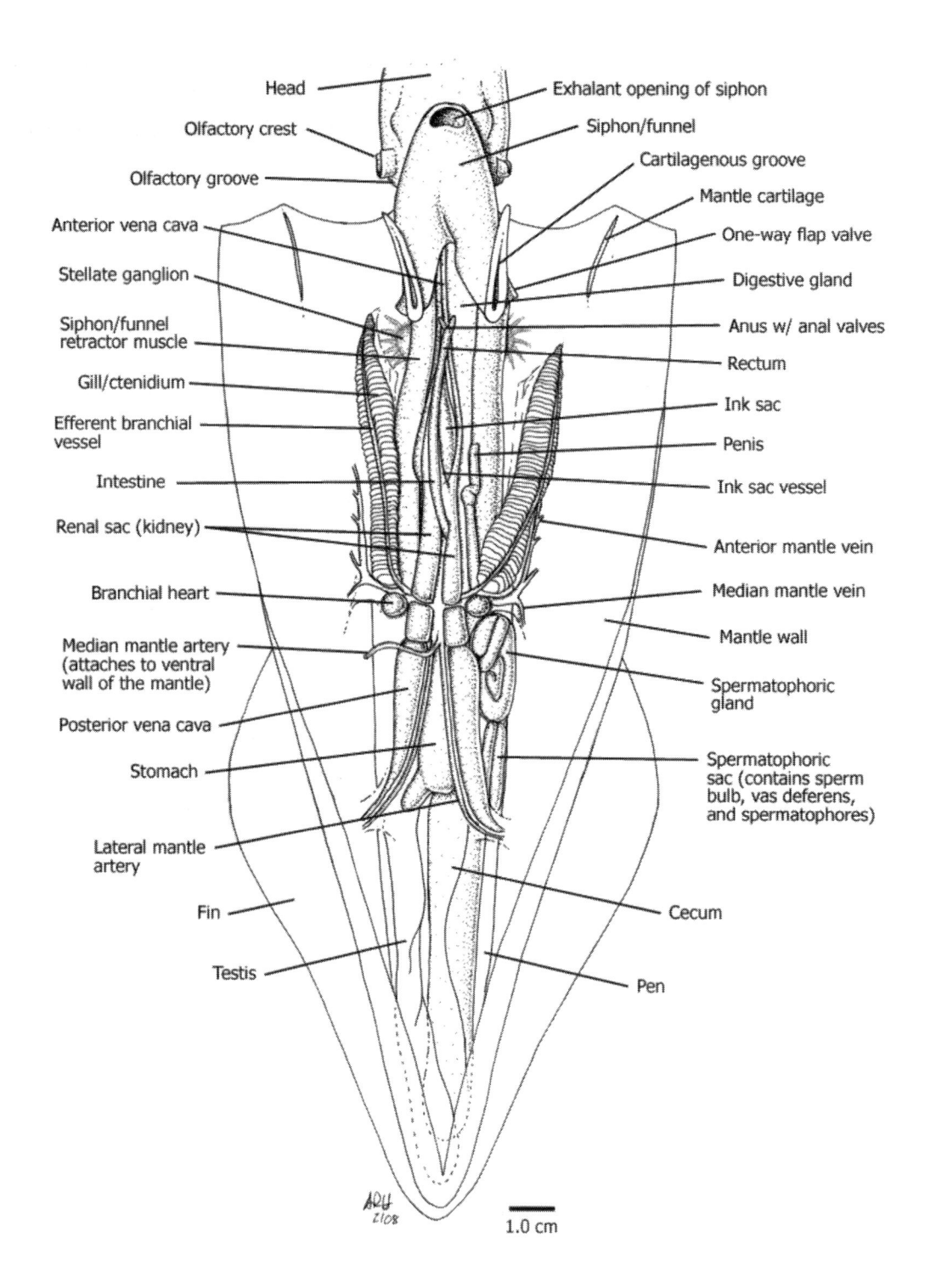

Figure 7.12. Structures of the visceral mass of a male squid, ventral view. (Image: ARH)

3) Determine the gender of your squid. If your squid is male you will see a cylindrical penis along the left side of the visceral mass between the left gill and the leading edge of the mantle (remember your squid is on its back). You should also see the coiled spermatophoric gland just posterior to the left gill. If your squid is female you will see the funnel-shaped opening of the oviduct on the left side of the visceral mass just anterior

to the left gill. There will also be a pair of large nidamental glands located between the branchial hearts and extending posteriorly from there. Refer to **Fig. 7.13** and **7.14** to help you determine the gender of your squid. It is sometimes difficult to sex squid so check with other lab groups to see what they can see in their squids if your squid's gender is not immediately obvious.

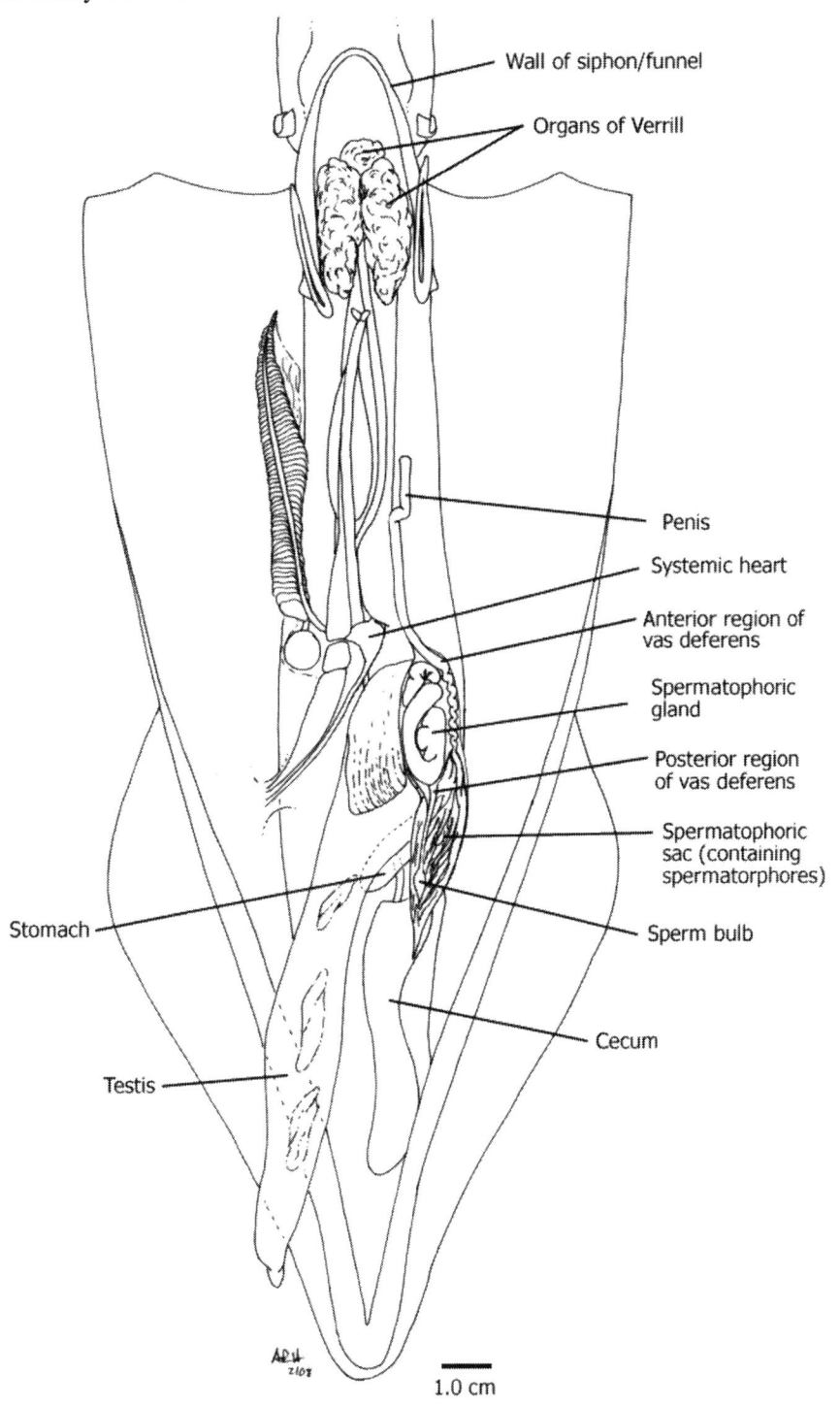

Figure 7.13. Anatomy of the male squid reproductive system, ventral view. (Image: ARH)

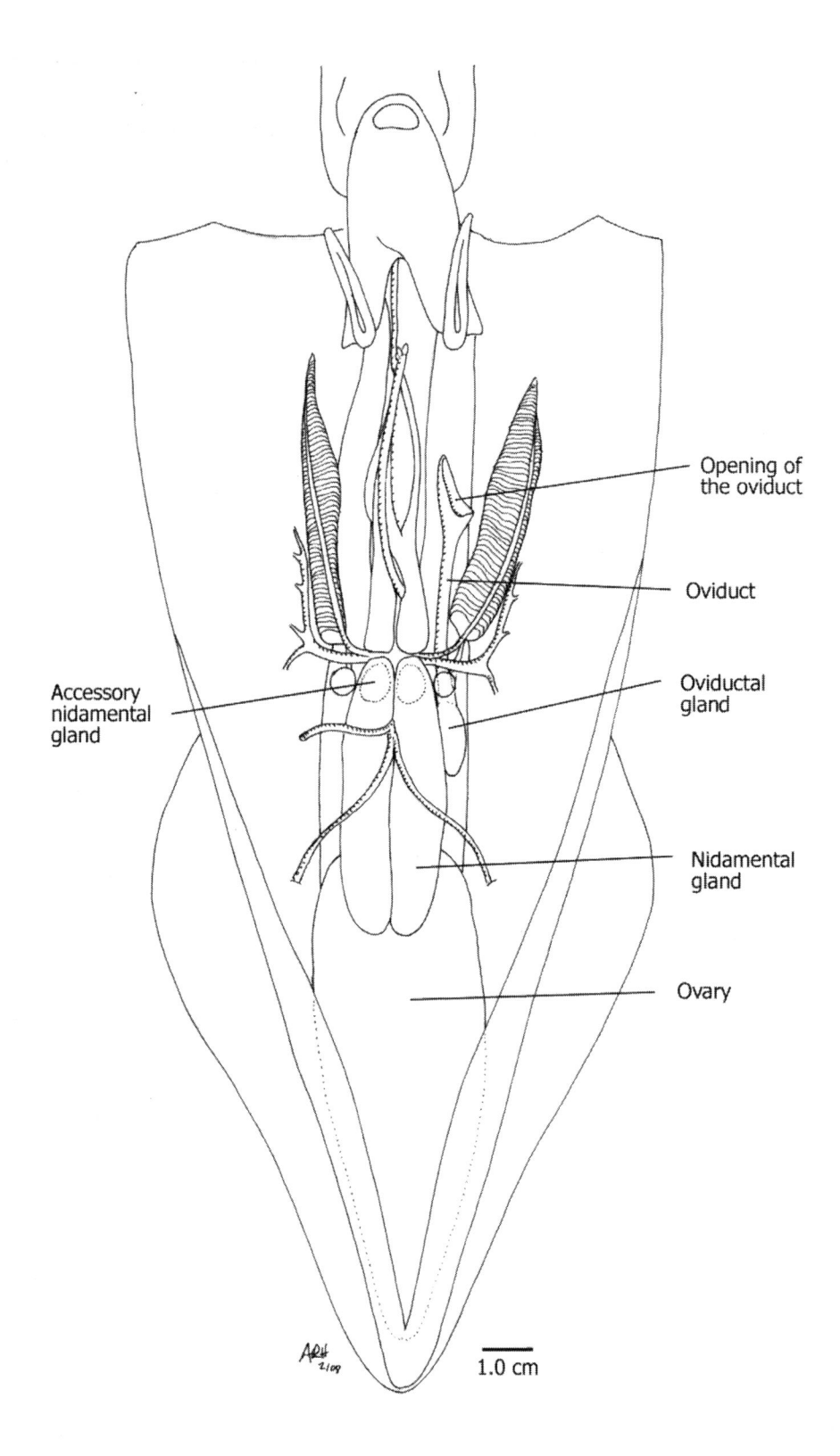

Figure 7.14. Female squid reproductive system, ventral view. (Image: ARH)

4) Male reproductive system. To see the entire male reproductive system carefully remove the following structures: the left ctenidium, left branchial heart, posterior vena cava and lateral mantle artery. Next remove the thin, transparent epithelial covering of the posterior portion of the visceral mass. DRAW the male reproductive system and refer to **Fig. 7.13** to help you identify what you see.
5) Female reproductive system. The female reproductive system is largely visible when you open the mantle cavity. Remove the left nidamental gland to expose the accessory nidamental gland underneath it and the oviductal gland and oviduct. DRAW the female reproductive system and refer to **Fig. 7.14** to help you identify what you see. If possible study both male and female squid. I recommend that teams teach each other the reproductive anatomy of their squid.
6) Circulatory system. Much of the circulatory system is visible when you open the mantle cavity, especially if your squid has been injected with colored latex. In order to see the systemic heart and circulatory structures of the gills you first need to carefully remove the kidneys. Examine the circulatory system and DRAW a figure that indicates the overall flow of blood through the body. This is particularly interesting since squid have 3 hearts.
7) Digestive system. Examine the digestive system only after you complete your investigations of all other anatomy of the mantle cavity. Remove the gills, branchial and systemic hearts, the entire reproductive system, the ink sac (be careful as you remove the ink sac, if it ruptures ink may leak all over your squid – set the ink sac aside in a small empty bowl), the ventral wall of the siphon and the siphon retractor muscles. Also separate the head retractor muscles from the anterior portion of the digestive gland. Be advised that the esophagus is long, delicate and easily broken. DRAW the digestive tract and refer to **Fig. 7.15** to help you identify what you see.
8) Observe the anatomy of the head, foot and buccal bulb. Examine the arms, tentacles and suckers. Compare the suckers of squid to those of octopus if octopi are available. Next, study the anatomy of the head and mouth. Expose the buccal bulb by cutting away the tissue around the eyes and the base of the arms. The buccal bulb is a mass of muscle and connective tissue that houses the beak, radula, mouth and opening to the esophagus and can move independently within the ring of arms. Find the esophagus coming out of the posterior end of the buccal bulb and then remove the buccal bulb from the head. Keep track of the dorso-ventral orientation of the buccal bulb. Make a cut completely through the midline of the buccal bulb and lay it open showing mirror halves. DRAW what you see (refer to **Fig. 7.16**).
9) If you have additional time study the nervous system. Carefully shave tissue off of the dorsal surface of the head until you find the cartilaginous cranium. Yes, squid have a cranium. Remove the cranium and expose the brain. Refer to your textbook (Brusca, *et al. 3e*, 2016, p. 510; Pechenik *7e*, 2015, p. 262; or Ruppert, *et al.*, 2004, p. 360) for help identifying parts of the brain. Next, carefully remove one of the eyes from your squid and open it. Examine the lens and the black pigment layer.
10) Lastly, a rite of passage for any student of invertebrate zoology. Open the ink sac. Dip a crow quill pen, the pen of your squid (the stiff chitinous structure found partially embedded in the dorsal wall of the mantle) or a probe tip into the ink and write your name in your lab notebook – add a few drops of water to the ink if needed. Note the brownish color of the ink. Cephalopods are historically a source of the brownish pigment "sepia."

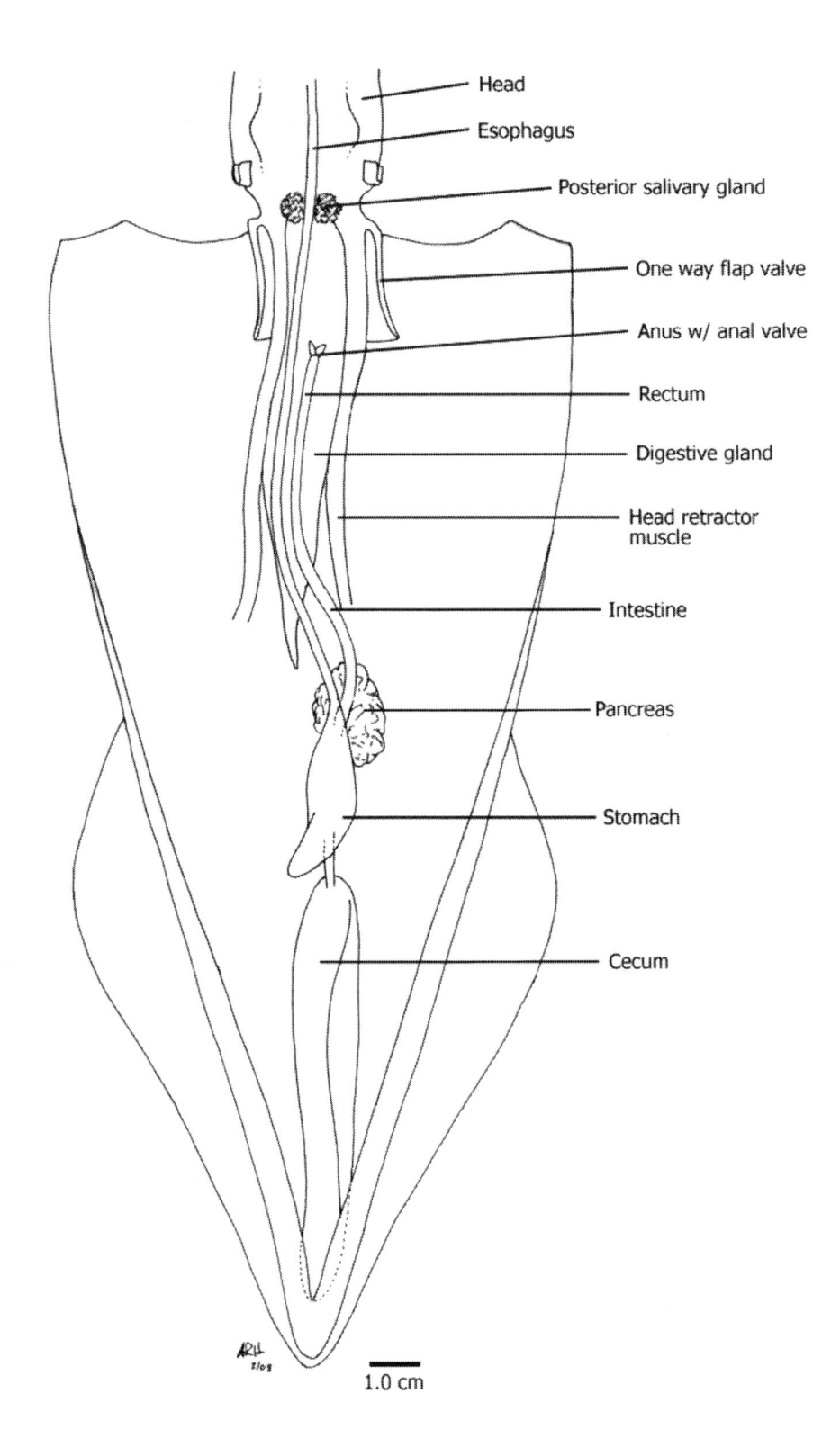

Figure 7.15. Squid digestive tract, ventral view. (Image: ARH)

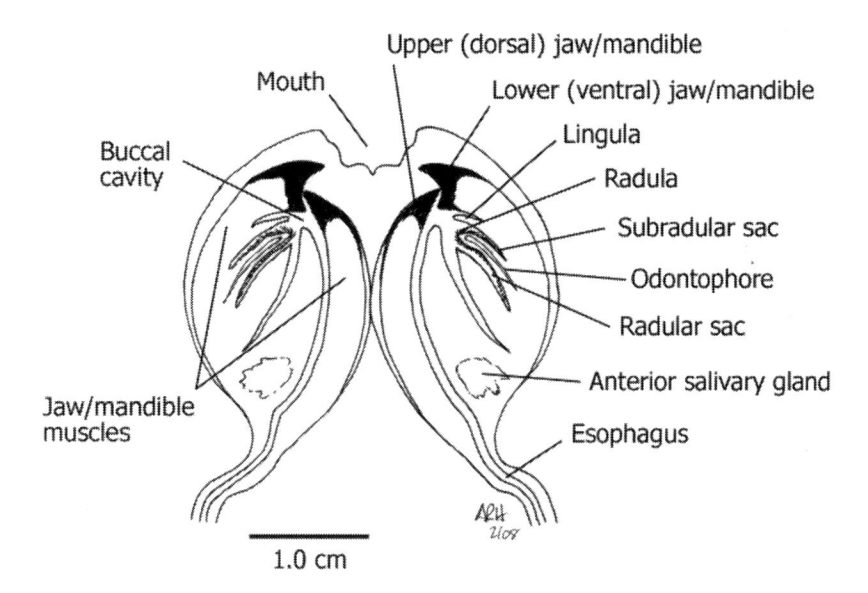

Figure 7.16. Squid buccal bulb. The ventral surfaces are toward the right and left edges of the drawing, and the dorsal surface is in the center of the drawing. (Image: ARH)

Class Bivalvia - clams

There are about 10,000 species of clams, mussels and their relations. Their bodies are highly derived; their soft tissues are housed in a pair of heavy shells and they have no discernable head. Most bivalves also use a large pair of ctenidia for gas exchange and suspension feeding. Bivalves live in nearly every aquatic habitat on earth, from wave-swept beaches and rocky intertidal communities to deep-sea hydrothermal vents and freshwater lakes, streams and rivers. The internal and external anatomy of bivalves shows some variability but once you learn the anatomy of one bivalve it is not particularly difficult to decipher the anatomy of any bivalve. The freshwater clam *Unio* is the focus of this exercise.

Tasks – Class Bivalvia

1) Examine the external anatomy of the shell. Pay particular attention to the shell hinge and ligaments. The siphonal region marks the posterior end of the animal. DRAW what you see and refer to **Fig. 7.17** to help you identify visible structures.
2) Open your specimen and examine the anatomy of inner surface of the left valve (shell). The best way to open the clam is to work the blade of a scalpel or a sharp knife between the valves of the shell to cut through the anterior and posterior adductor muscles. **Figure 7.18** shows the location of these muscles. You will know when they have been cut because you will be able to separate the valves of the shell with relative ease. You will also know when you are cutting these muscles because you will be able to hear and feel a distinctive snapping or cracking sound as individual muscle fibers are cut. Before you pull the shells apart, use the tip of a probe to carefully scrape the inner surface of the shell and separate all soft tissues from the inner surface of the shell. Once all soft tissues are separated from the shell open the clam so that the soft tissues rest in the right valve. DRAW the inner surface of the left valve and refer to **Fig. 7.18** to help you identify what you see. Take time to examine the internal and external portions of the hinge.

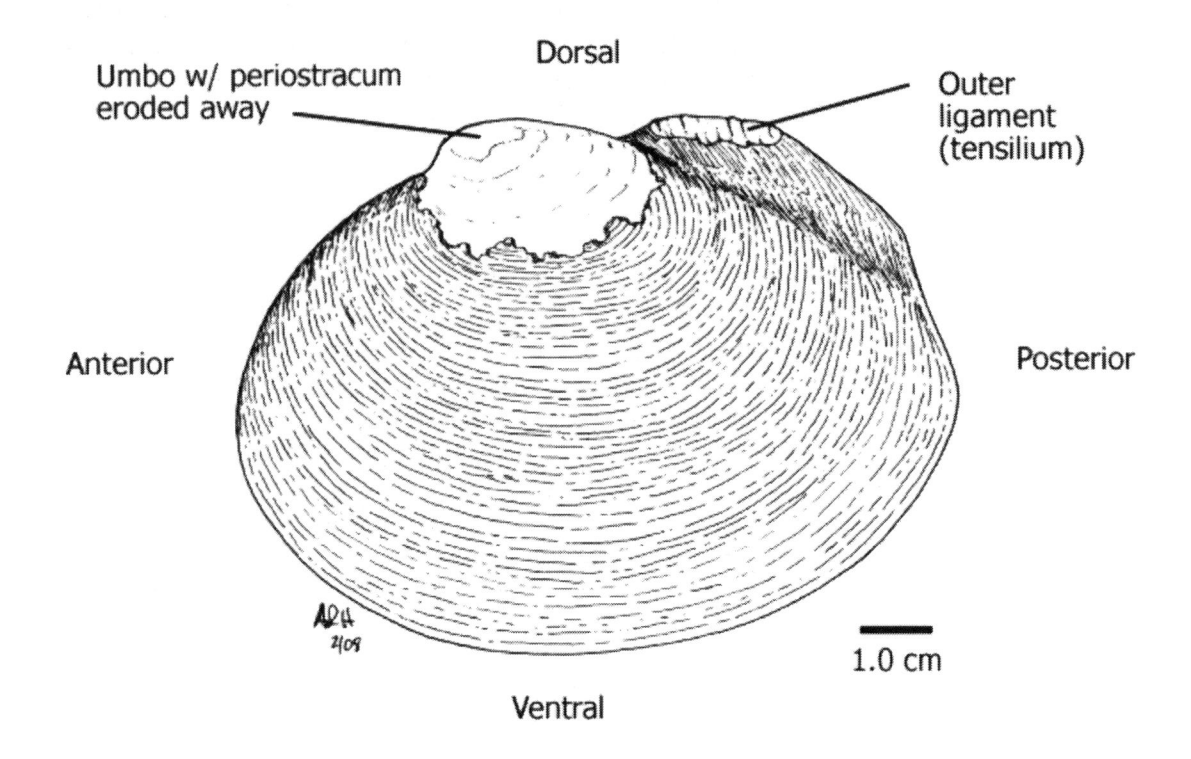

Figure 7.17. External anatomy of the left valve (shell) of *Unio*. (Image: ARH)

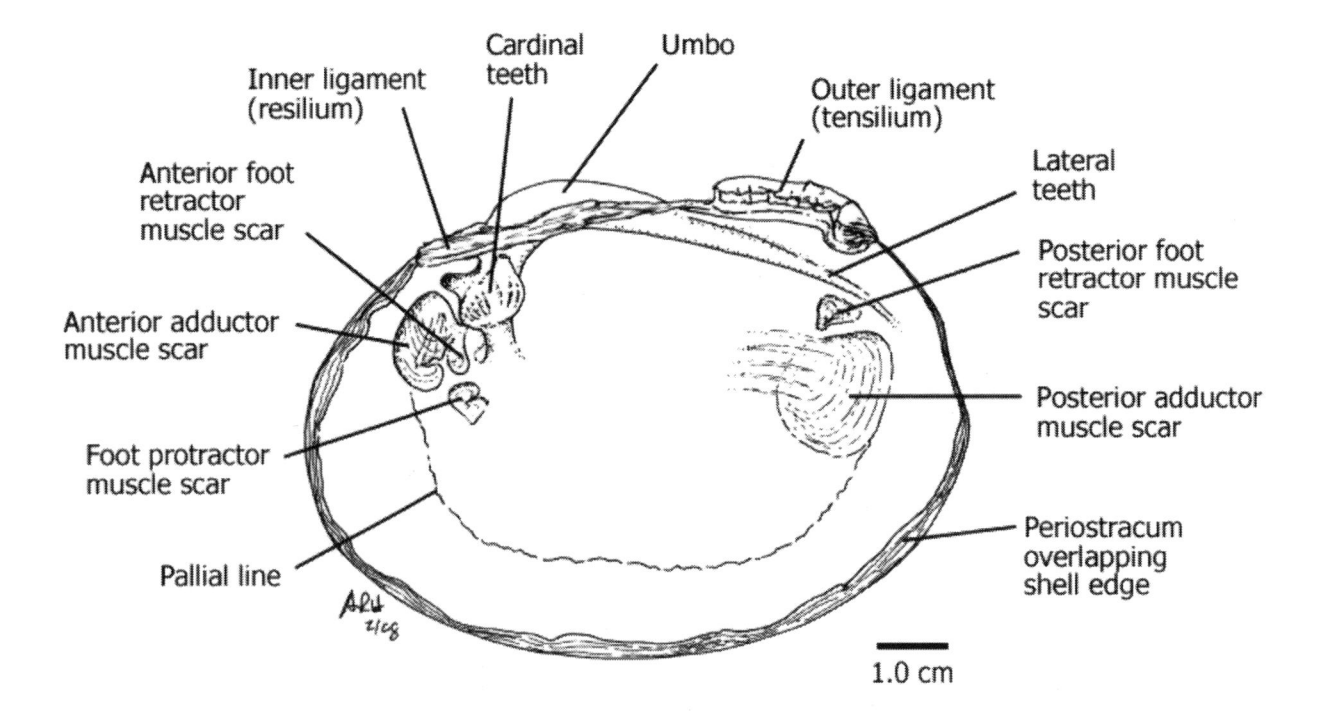

Figure 7.18. Internal shell anatomy of right valve of *Unio*. (Image: ARH)

3) Immerse your clam and observe the anatomy of the exposed soft-tissue. DRAW the soft tissue structures you can see. Refer to **Fig. 7.19** to help you identify exposed anatomy. Look at the posterior end of your clam and identify incurrent and excurrent siphonal regions. Refer to **Fig. 7.20** to help you identify what you see.

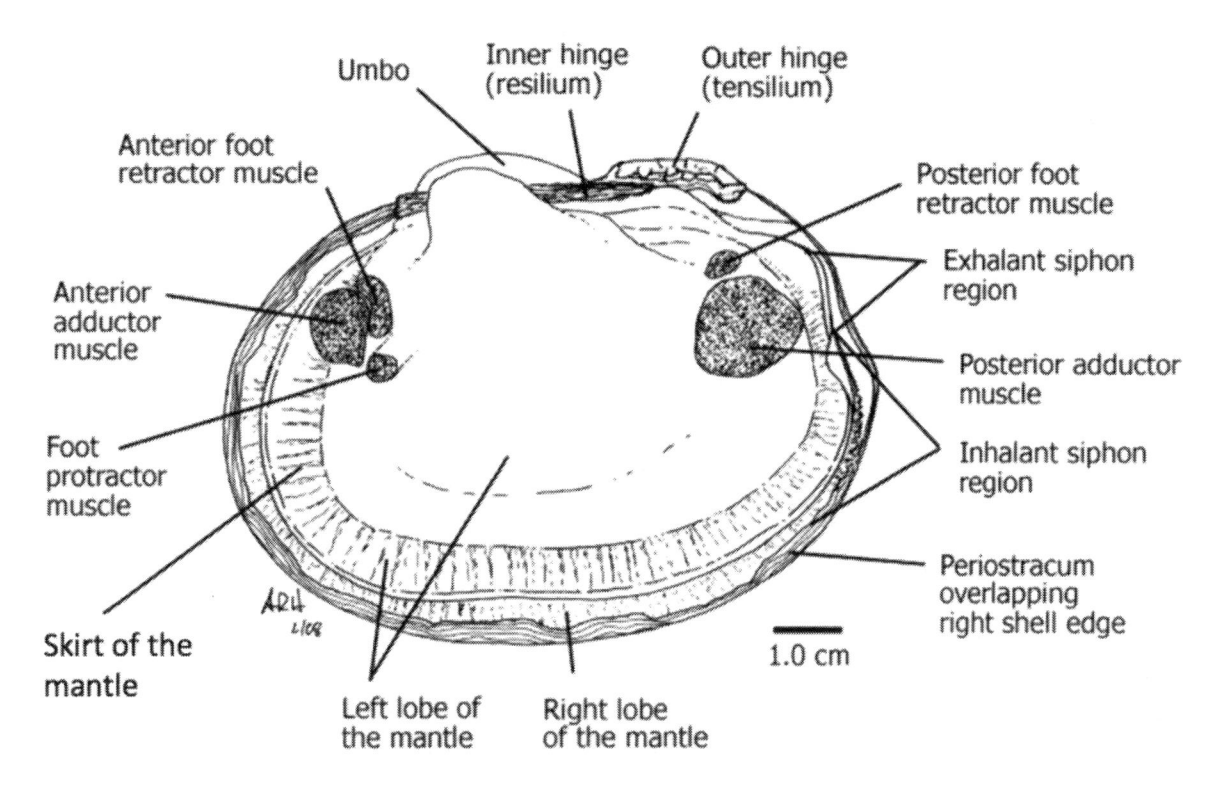

Figure 7.19. Soft tissues of *Unio* revealed by removing the left valve of the shell. (Image: ARH)

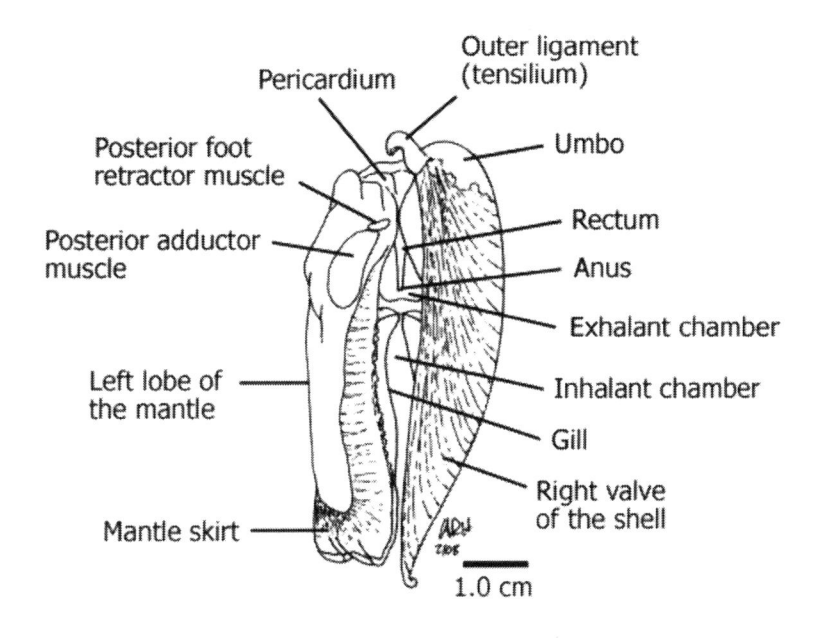

Figure 7.20. Posterior view of *Unio* with the left valve removed. (Image: ARH)

4) Remove the left lobe of the mantle but do not to cut into the pericardium or nephridial sac. DRAW what you see and refer to **Fig. 7.21** to help you identify exposed structures.

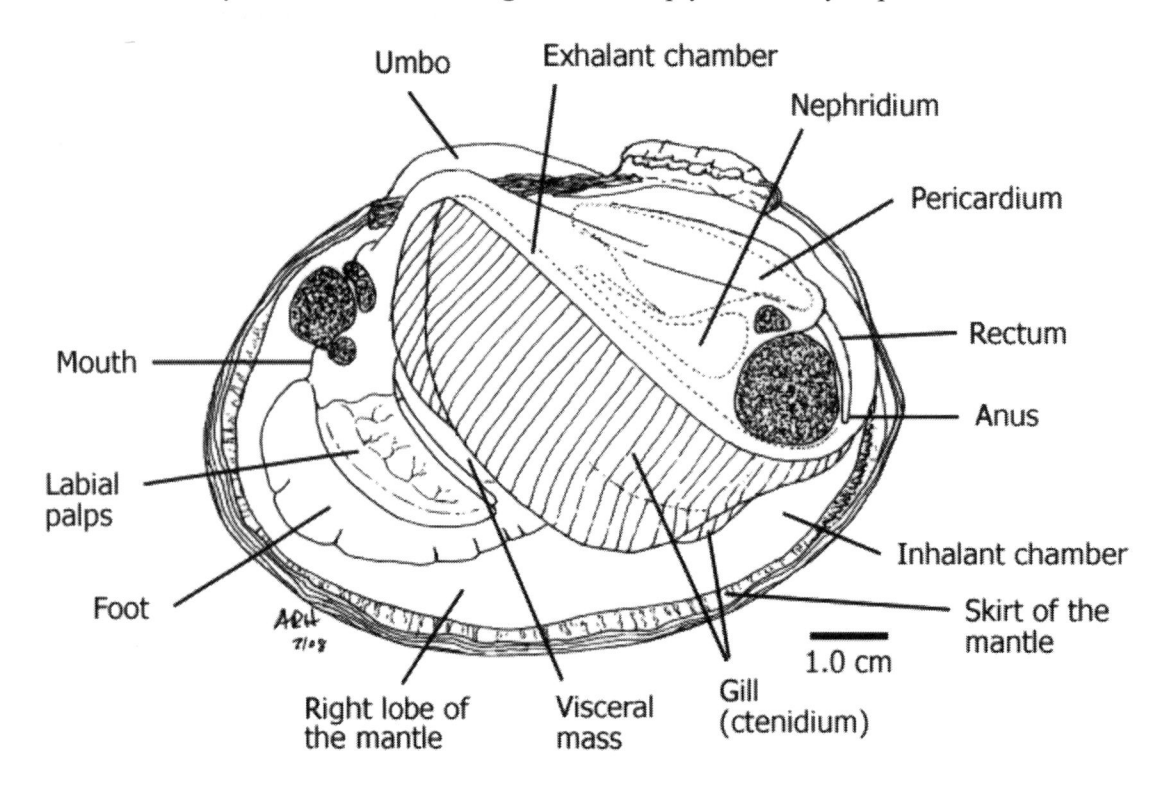

Figure 7.21. Structures of the left mantle cavity of *Unio*. (Image: ARH)

5) Remove the left ctenidium (gill) and carefully shave away layers of tissue and muscles on the left side of the visceral mass to expose dull yellow and olive green masses of tissue within. There is also a small, coiled tube running through the visceral mass that will be evidenced by small holes in the digestive gland and ovary. These holes are the lumen of the stomach and intestine. The digestive gland is green and the gonads are yellow. Find the mouth. It is located on the anterior midline of the body directly between the left and right sets of labial palps. Insert a blunt probe into the mouth and use it to locate the stomach. Next, open the pericardium and nephridial sac. DRAW what you see and refer to **Fig. 7.22** to help you identify these structures.
6) Examine a prepared slide of glochidia larvae of freshwater bivalves. DRAW one glochidia larva and refer to **Fig. 7.23** to help you identify what you see.

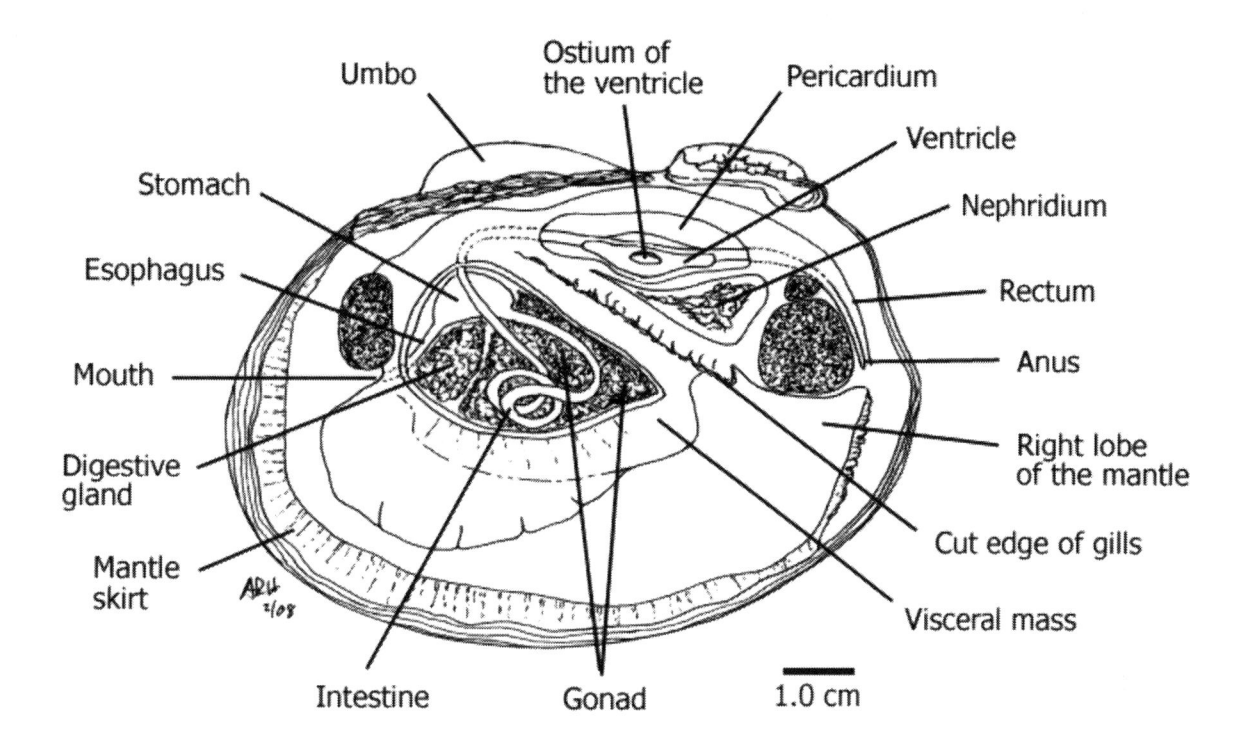

Figure 7.22. Internal anatomy of *Unio*. (Image: ARH)

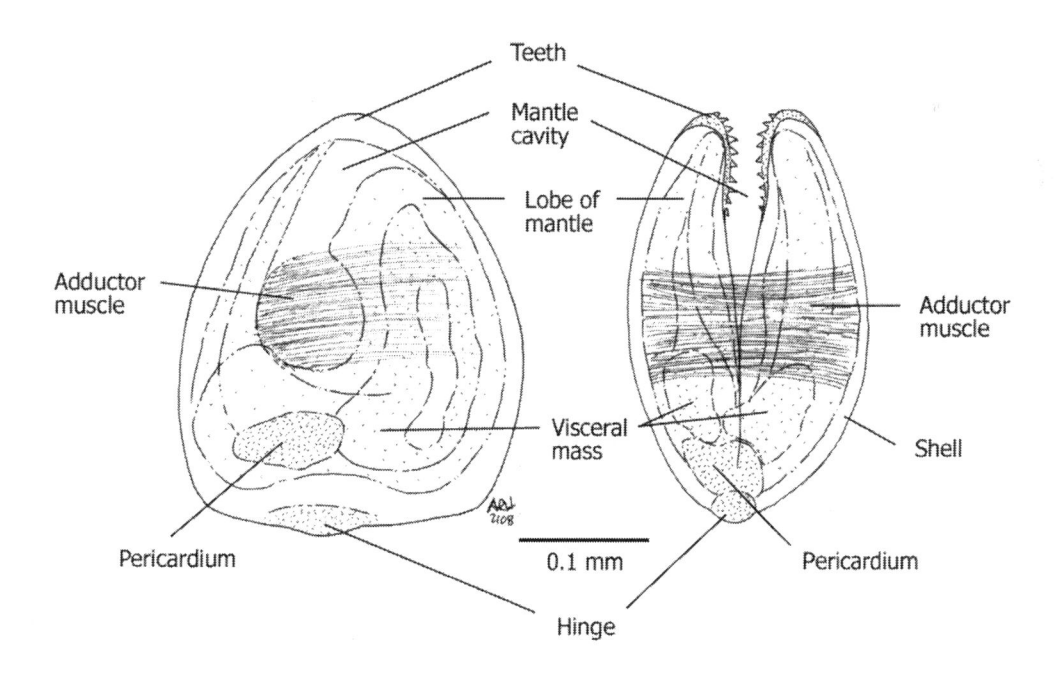

Figure 7.23. Glochidia larva of a freshwater mussel. Watch this video to see how freshwater mussels are able to get their parasitic young, the glochidia, into the mouths and gills of fish: https://www.youtube.com/watch?v=I0YTBj0WHkU. (Image: ARH)

Group Questions

1) How does the proportion of the body allocated to the digestive system in molluscs compare to what you saw in the body plans of the Platyhelminthes? Develop a hypothesis that explains the differences in allocation of body space to digestion in Molluscs and Platyhelminthes.
2) Develop a hypothesis that explains how the basic body plan of molluscs allowed them to become so diverse as well as ecologically and evolutionarily successful.
3) Develop a hypothesis that explains why freshwater bivalves have glochidia larvae while marine bivalves do not.

Phylum Mollusca – Glossary

Aesthetes – shell canals/light sensory structures in chitons

Adductor muscle – powerful muscle used to close the shells of bivalves

Adductor muscle scar – location where an adductor muscle attaches to the shell

Afferent branchial vessel – carries deoxygenated blood from the branchial heart to the ctenidia

Albumin gland – organ that produces yolk

Anterior foot retractor muscle – muscle that pulls the foot in the anterior direction

Anterior foot retractor muscle scar – location where this muscle attaches to the shell

Anterior mantle vein – collects deoxygenated blood from the anterior portion of the mantle and carries it to the branchial heart

Anterior salivary gland – secretes saliva that helps lubricate food as it moves through the esophagus

Anterior vena cava – vessel that collects deoxygenated blood from the anterior part of the body

Aorta – major blood vessel leaving the heart, carries hemolymph to blood sinuses

Aplacophora – small wormlike molluscs that do not produce a shell

Aquiferous pore – opening near the cephalopod eye, believed to be chemosensory or for equalizing pressure

Arms of cephalopods – muscular sucker-bearing appendages used for location and prey capture

Branchial heart – heart that pumps blood through the afferent branchial artery to the ctenidia

Buccal bulb – organ housing the mouth, buccal cavity, radula/odontophore complex and salivary glands in squid

Buccal cavity – space inside the mouth anterior to the radula

Captacula – threadlike ciliated feeding tentacles of scaphopods

Cardinal teeth – large interlocking extensions of bivalve shells that aid in shell articulation and oppose sheer stress

Cartilaginous groove/mantle cartilage – the mantle cartilage fits into the cartilaginous groove in a lock-and-key arrangement when circular muscles surrounding the aperture of the mantle cavity contract and they help prevent water from escaping anywhere besides the siphon during jet propulsion by squid

Cecum – organ that stores and sorts food particles and determines which particles go to the digestive gland for digestion and which are sent to the intestine for elimination from the body

Coelomate body – body plan that includes a fluid-filled space that is lined completely by mesodermally derived peritoneum

Collar – thick rim of tissue that lines the margin of the mantle cavity and shell aperture in terrestrial snails

Common genital opening – the opening for both the female and male reproductive systems

Copulatory bursa – small organ that secretes lubricant to help sperm/spermatophores move more easily during reproduction

Crop – sac-like organ located between the esophagus and stomach, salivary glands are located on the lateral surfaces of the crop in *Helix*

Ctenidia – mollusc gills

Dart sac – organ that produces darts of $CaCO_3$ or cartilage, darts may be 1-30mm long depending on the species and are pushed out through the common genital opening like a lance into a prospective mate's foot perhaps repeatedly, and this appears to trigger or intensify mating attempts eventually resulting in simultaneous exchange of spermatophores

Detorsion – developmental process that reverses the effects of torsion

Digestive gland – organ where most of the extracellular digestion and food absorption occurs

Efferent branchial vessel – carries oxygenated blood from the ctenidia to the systemic heart

Esophagus – section of the digestive tract between the pharynx and the stomach

Exhalant aperture – region of the posterior girdle of chitons where water is released from the mantle cavity

Exhalant chamber of the mantle cavity – space in the mantle cavity for water that has already passed across the ctenidia

Exhalant siphon region – section of the mantle through which water is released from the mantle cavity

Fin – lateral extension of the squid mantle, provides stability during rapid swimming

Flagellum – organ that contributes to the production of spermatophores in gastropods

Foot protractor muscle – muscle that extends the foot

Foot protractor muscle scar - location where the foot protractor muscle attaches to the shell

Girdle – tough fleshy tissue in chitons that anchors the eight shell plates and protects the margin of the body and organs of the mantle cavity

Glochidia larva – larval stage of freshwater mussels that are parasitic on gills and fins of fishes

Gonopore – opening through which gametes are released

Head (cephalic) retractor muscles – large muscles that pull the head and margin of the mantle of squid together during jet propulsion

Inhalant aperture of the mantle cavity – location where water is pulled into the mantle cavity

Inhalant chamber of the mantle cavity – space in the mantle cavity where water has not yet passed across the ctenidia

Inhalant siphon region – section of the mantle through which water is pulled into the mantle cavity

Ink sac – organ that produces ink, ink is released directly into the cephalopod siphon allowing it to be quickly and effectively ejected into the water

Inner ligament (resilium) – tough proteinaceous structure that is compressed when the valves of a bivalve shell are closed and pushes the shells apart when adductor muscles relax

Intestine – tubular organ that carries nondigestible material and waste material to the anus

Jaw/mandible – beak-like organ of squid used to penetrate the body wall and tear prey into small pieces that can be processed by the radula before swallowing

Labial palps – pairs of ciliated muscular flaps that flank the mouth in bivalves, they collect food from the ctenidia and move it to the mouth

Lateral mantle artery – carries oxygenated blood from the systemic heart to the mantle

Lateral teeth – a series of a few to many small interlocking teeth that help prevent the two shells (valves) of bivalves from slipping sideways due to shear stress

Limpet – gastropod that produces a non-coiled, cap-shaped shell

Lingula – presumably a chemosensory/taste organ

Mantle - tissue layer that secretes the shell and lines the inner surfaces of shells or may be thick and muscular as in cephalopods

Mantle cavity – fluid-filled cavity that houses the ctenidia and other organs

Mantle skirt – portion of the mantle that extends beyond the pallial line, this skirt allows the animal to secrete new shell at shell edges and to withdraw the mantle when needed

Median mantle artery – vessel that carries oxygenated blood from the systemic heart to the tissues of the ventral mantle

Median mantle vein – collects deoxygenated blood from the medial region of the mantle and carries it to the branchial heart

Metanephridium – excretory organ with a cilia-lined incurrent opening, a tubule where primary urine is modified by absorption and secretion and a nephridiopore where final urine is released

Nephridiopore – opening though which nitrogenous waste (urine) is released

Nerve ring – fused cerebral ganglia that form a ring of nerve tissue around the esophagus

Nidamental glands and accessory nidamental glands – produce protective casings for individual eggs

Odontophore – cartilaginous structure that supports the radula

Olfactory crest and groove – Chemosensory organ located on both sides of the head in squid

One-way valve – allows water to move in only one direction through the siphon

Organs of Verrill – located inside the siphon, function unknown

Ostium of the ventricle – opening through which hemolymph is pulled into the ventricle from the pericardium

Outer ligament (tensilium) – a tough proteinaceous structure that is stretched when the valves of a bivalve shell are closed, tension from this ligament causes shells to gape when adductor muscles relax

Oviduct – tube that carries eggs from the ovary to the gonopore

Oviductal gland – organ that secretes outer coverings of egg sacs, each sac can contain many eggs

Ovo/testis – reproductive organs that produce eggs and sperm

Pancreas – organ that secretes digestive enzymes

Pericardium – fluid-filled sac surrounding the heart

Pneumostome – literally means "air mouth" and is the opening in terrestrial gastropods where gas is pulled into the mantle cavity and exhaled from the mantle cavity

Posterior foot retractor muscle – muscle that pulls the foot in a posterior direction

Posterior foot retractor muscle scar – location where this muscle attaches to the shell

Posterior salivary gland – an organ in squid that secretes toxins and the digestive enzyme hyaluronidase, a substance that breaks down hyaluronic acid (a component of connective tissues)

Posterior vena cava – vessel that collects deoxygenated blood from the posterior portion of the body

Radula – tooth-bearing tongue-like strap used for scraping up or rasping on food items during feeding

Radular sac – cavity where the radula is constantly secreted

Rectum – terminal segment of the intestine that carries feces to the anus

Renal sac (kidney) - organ that removes nitrogenous waste from the blood

Salivary glands – organs that secrete mucus that helps food move smoothly through the digestive system, saliva may include digestive enzymes or toxins

Shell extensor muscles of chitons – muscles attached to the dorsal surfaces of shell plates that cause chitons to unroll

Shell flexor muscles of chitons – muscles attached to the ventral surfaces of shell plates that cause chitons to roll up

Siphon/funnel (squid) – muscular organ through which water is forcibly expelled to produce jet propulsion, the siphon is fully directional allowing squid to jet in any direction

Siphon/funnel retractor muscle – muscle used for directional control of the siphon

Slug – elongate gastropod that lacks a shell and undergoes detorsion

Sperm bulb – collects sperm from the testis and moves it into the vas deferens

Spermatophoric sac (Needham sac) – sac where spermatophores are stored prior to reproduction

Spermatophoric gland – organ where spermatophores are produced

Stellate ganglion – large nerve ganglion located dorso-lateral to the siphon of squid, where giant squid axons synapse and gives rise to neurons that innervate tissues of the mantle

Stomach – saclike organ where food is stored and sorted and is where digestion begins

Subradular sac – space ventral to the radula/odontophore complex where old radula material is constantly resorbed

Suspension feeding – feeding strategy where particles floating in the water are collected, sorted and ingested

Systemic heart – contractile organ that receives oxygenated blood from the ctenidia and pumps it to the body

Tentacles (of cephalopods) – long appendages with terminal sucker-bearing pads, can be rapidly extended during prey capture

Tentacles (of gastropods) – non-segmented tactile, chemosensory or photo-sensory structures that extend from the head, foot or mantle. **Note**: Do not confuse tentacles with antennae. Antennae are always segmented or jointed and are produced by arthropods and their relations.

Torsion – 90^0 to 180^0 rotation of the visceral mass and shell relative to the head-foot in gastropods, contractions of larval retractor muscles and differential tissue growth in the veliger stage results produce a U-shaped gut and twisting of longitudinal nerve cords that extend into the visceral mass, also moves the mantle cavity into an anterior position

Trochophore larva – larval stage produced by most mollusc taxa, has an oblong body with a dome-shaped upper portion, two bands of cilia running around the equator that are used for locomotion and feeding and tufts of cilia at the dorsal and ventral poles

Umbo – oldest point on a bivalve shell

Vagina – structure that receives sperm or spermatophores during mating

Vas deferens – tube that carries spermatophores from the testis to the penis

Ventricle – contractile chamber of the heart

Visceral hump – Blind-ended tube or sac surrounded by the mantle in cephalopods, houses organs of the mantle cavity and the visceral mass

Visceral mass – the soft non-muscular region of the mollusc body that houses the internal organs, is located dorsal to the head/foot and is covered and protected by the mantle and shell

Chapter 8: Phylum Annelida

Phylum Annelida includes about 20,000 species including segmented worms and their relatives. Annelida means "little rings" and refers to the ring-like appearance of segments in the traditional annelid body plan.

The taxonomy of this phylum has experienced some interesting changes. The annelids used to include only earthworms and their relations, polychaete worms and their relations and leeches. Molecular phylogenetic research has shown however that other groups of animals also belong in this phylum. Characteristics of annelids are listed in **Table 8.1**.

Table 8.1. Characteristics of Phylum Annelida (after Brusca, *et al.*, 2016).

- Worms that are segmented (in most), triploblastic, coelomate and bilaterally symmetrical
- Metamerism (in most)
- Chaetae (in most)
- Metanephridia
- Closed circulatory system with hemoglobin, hemocyanin or chlorocruorin blood pigment
- Regional specialization of a complete digestive system
- Trochophore larva (in some)

Phylum Annelida "little rings" - Taxonomy

As mentioned above, the taxonomy of Annelids has undergone significant changes. Phylogenetic analysis has resolved many long-standing questions about a variety of groups of animals. This work has shown that some taxa that were previously recognized as independent phyla are actually derived annelids. These include sipunculans (peanut worms), echiurans (spoon worms), and pogonophorans and vestimentiferans (two groups of beard worms).

The current taxonomy of the annelids is too complex to treat adequately here; only selected taxa are described below.

>**Family Sipuncula "small tube"** (covered in this exercise) – peanut worms. Sipunculans were formerly identified as a phylum but they are highly derived annelids that are now placed among other basal annelid taxa. Sipunculids have lost many annelid characteristics and evolved unique ones. These worms have a retractable proboscis and a U-shaped gut, but they lack segmentation, a closed circulatory system and chaetae. They also have a highly unusual epidermis that has longitudinal coelomic channels not seen in other animals. This video shows a sipunculid worm that has been dug up and placed on the sediment surface. Normally the trunk is in the substrate or hidden under a rock but in this video you can see the entire body plan including trunk and eversible introvert: https://www.youtube.com/watch?v=WA5B3ps2GAw

Subclass Errantia "wanderers" – These are mainly polychaete worms with parapodia and many long chaetae, i.e., polychaete worms.

> **Family Nereididae "sea nymph like"** (covered in this exercise) – these are well-known polychaete worms. They are active predators or scavengers in intertidal and shallow marine habitats including estuaries, though some deep-water species exist. This video shows a live nereid worm. Be sure to watch long enough to see it extend its curved jaws: https://www.youtube.com/watch?v=LdHzRNdz2HU

Subclass Sedentaria "sitting still" – this is a clade that contains many tube-dwelling worms as well as the worms that produce a clitellum such as earthworms and leeches.

> **Family Siboglinidae "water serpent moving back and forth"** (not covered in this exercise) – beard worms. We have known about these animals since the 1800s. They have long thin bodies but only the posterior section of the body shows evidence of segmentation and bears a single ring of chaetae – this is the part of the body that inevitably breaks off and is left in the substrate while collecting specimens via bottom dredging. The rest of the body did not show any anatomical similarities to annelids and their taxonomic affinity was not resolved until much more recently when undamaged specimens could be retrieved and studied in detail. Hydrothermal vent communities were discovered in the late 1970s and two-meter long vestimentiferan beard worms made biological news. We originally didn't know what to make of these worms either. They lacked a mouth and a functional gut and instead had a large internal compartment that was packed with chemosynthetic bacteria. Beard worms were originally regarded as a unique phylum but subsequent analyses shows that they are also derived annelids. The vent worms in this video belong to this family and the genus *Riftia*: https://www.youtube.com/watch?v=2FFnrW_SUdM

> **Family Echiuridae "viper tail"** (not covered in this exercise) – spoon worms. These highly derived annelids were also formerly designated as a phylum but molecular phylogenetics showed that they are derived annelids. These worms have an extensible proboscis that may be short or many times the length of the body, depending on the species. Echiurans have a single pair of chaetae anteriorly and a ring a chaetae around the anus. The proboscis is used by most species to carry out deposit feeding but a few species including *Urechis caupo*, the fat innkeeper worm, uses it to secrete a mucous net for suspension feeding. This animated video from the New York Times (online) does a nice job of depicting the unique reproductive biology of the echiuran worm *Bonellia viridis*: https://www.youtube.com/watch?v=v-MQxYFEHJo

> **Family Clitellata "with a saddle"** (covered in this exercise) – earthworms, leeches and relations. These annelids produce a clitellum that is used to make a cocoon to protect their offspring until they hatch; development is direct. These worms lack parapodia, have reduced or missing chaetae and are hermaphroditic. This video clip shows a giant earthworm and its cocoon: https://www.youtube.com/watch?v=uO4lkv-jLRs

Family Sipuncula – peanut worms

Sipunculans differ from other annelids in that they have a J-shaped gut and an anterior retractable structure called the introvert and they lack a segmented body and chaetae. They are similar to other annelids in that they have trochophore-like larvae, a cross-linked cuticle made of collagenous fibers, a similar body wall and nuchal organs.

Sipunculans are strictly marine organisms, have a worm-shaped body and are commonly found living in cracks, among cobbles or as infauna in soft-sediment environments.

Tasks – Family Sipuncula

1) Examine the external anatomy of your specimen. DRAW what you see.
2) Dissection – internal anatomy.
    a. Immerse your specimen and make as incision that runs the entire length of the body. There are few obvious indicators of this animal's dorso-ventral polarity but you will hopefully get lucky and be able to see paired structures such as the introvert retractor muscles and metanephridia after you open the body wall.
    b. Note the prominent longitudinal muscle bands that run the length of the body, and muscles associated with the intestine – the spindle muscle and radial fixing muscles.
    c. Look for the longitudinal nerve cord. It is lighter in color than the muscle bands of the body wall. Take particular care to study the anterior portion of the nerve cord and its relationship to the introvert and anterior region of the gut.
    d. There are two large brownish sac-like structures just posterior to the introvert. These are metanephridia. The gonads are easily detached from the body wall during initial stages of dissection but if you are careful you may see them attached to the body wall near the base of one of the introvert retractor muscles.
    e. DRAW what you see and refer to **Fig. 8.1** to help you identify visible structures.

Figure 8.1. Internal anatomy of *Sipunculus nudis*. (Image: ARH)

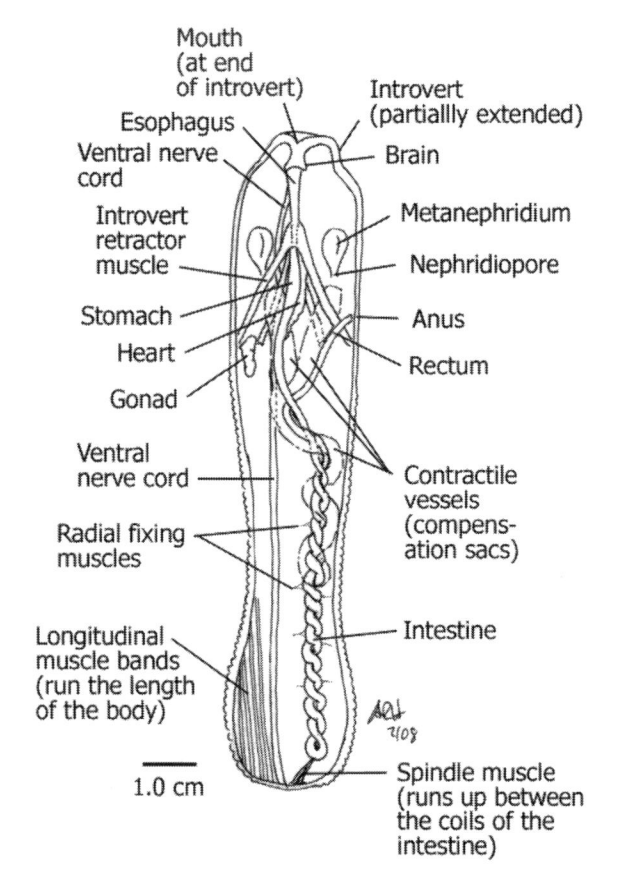

Subclass Errantia, Family Nereididae – polychaete worms

This exercise focuses on the common polychaete worm *Nereis*. It is an active hunter that has strong grasping jaws and a large muscular pharynx. *Nereis* is an osmoconformer and can survive in water of variable salinities from full-strength seawater to brackish water of estuaries.

Tasks - *Nereis*

1) Examine the head of *Nereis*. This worm identifies its food and senses its environment mainly via chemical and tactile stimuli. Structures of the head reflect this life style. Use a magnifying lens or dissecting scope to examine your specimen. DRAW the head of your specimen and refer to **Fig. 8.2** to help you identify what you see.

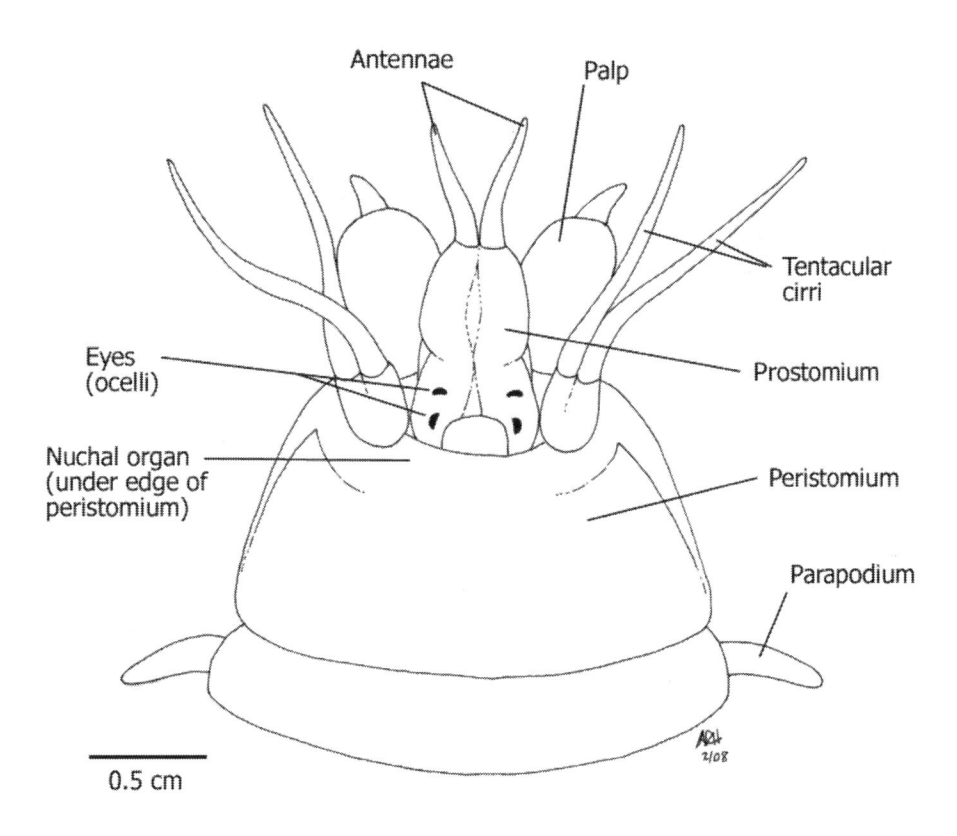

Figure 8.2. Structures of the head of *Nereis*, dorsal view. (Image: ARH)

2) Examine parapodia. Use a magnifying glass or dissection scope to examine the parapodia associated with a body segment of a preserved specimen of *Nereis*. Use a compound scope to examine a prepared slide of a parapodium of *Nereis*. DRAW what you see. Refer to **Fig. 8.3** to help you identify structures of parapodia.

3) Internal anatomy. Put a preserved specimen of *Nereis* in a wax bottomed dissection tray and add enough water to cover the worm. Starting at least 30 body segments back from the head, make a longitudinal incision through dorsal body wall, tease the body wall back and use insect pins to attach the body wall to the dissection tray. The coelomic spaces of all segments are separated from each other by transverse septae. Continue cutting and pinning the body wall as you move toward the head until you have exposed the internal

anatomy of the entire anterior region of your specimen. DRAW what you see and use **Fig. 8.4** to help you identify the anatomy of your specimen.

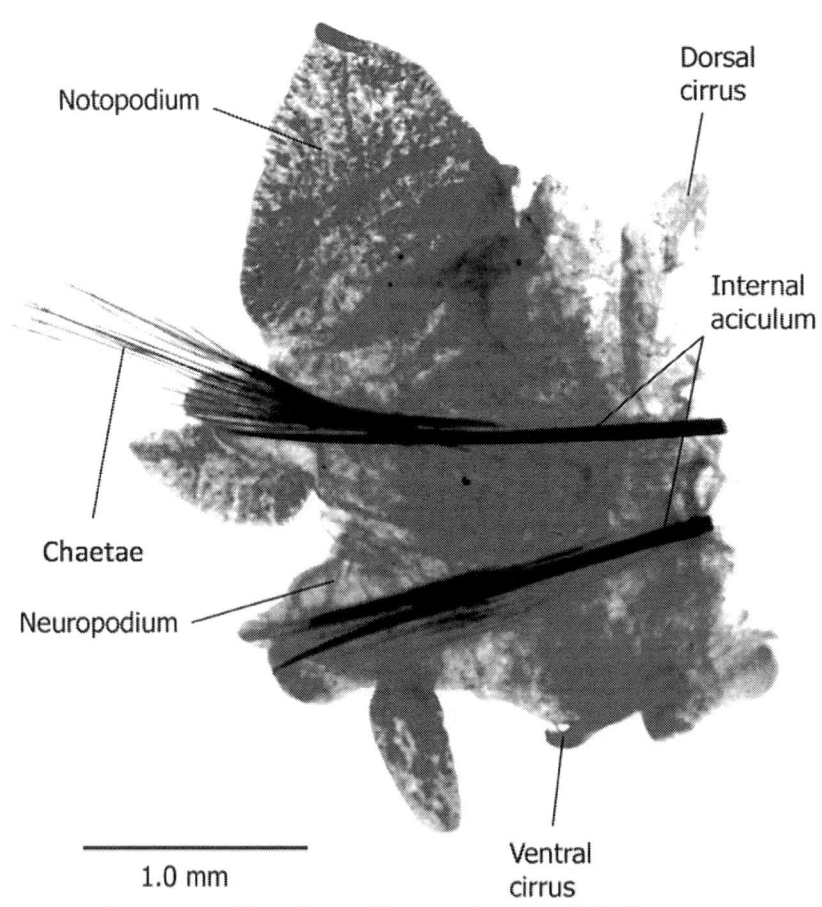

Figure 8.3. Anatomy of a parapodium from *Nereis*. (Image: ARH)

4) Dissect the pharynx. Look for a mass of muscles near the mouth. Open this structure and look for the pincer-like jaws. These are hook-shaped and are normally black in color. DRAW what you see.
5) Remove a section of the intestine so you can see the ventral blood vessel (dark tube/line) and the ventral nerve cord (light line). Add these structures to your drawing.
6) Look for a pair of metanephridia housed in the lateral areas of each body segment. Use a magnifying lens or dissection scope to find them. Nephridia can be delicate and are often destroyed when the body wall is opened. DRAW a metanephridium if you are able to locate one.
7) Use a compound scope to examine a prepared cross-section slide of *Nereis*. DRAW what you see and use **Fig. 8.5** to help you identify the anatomy of your specimen.

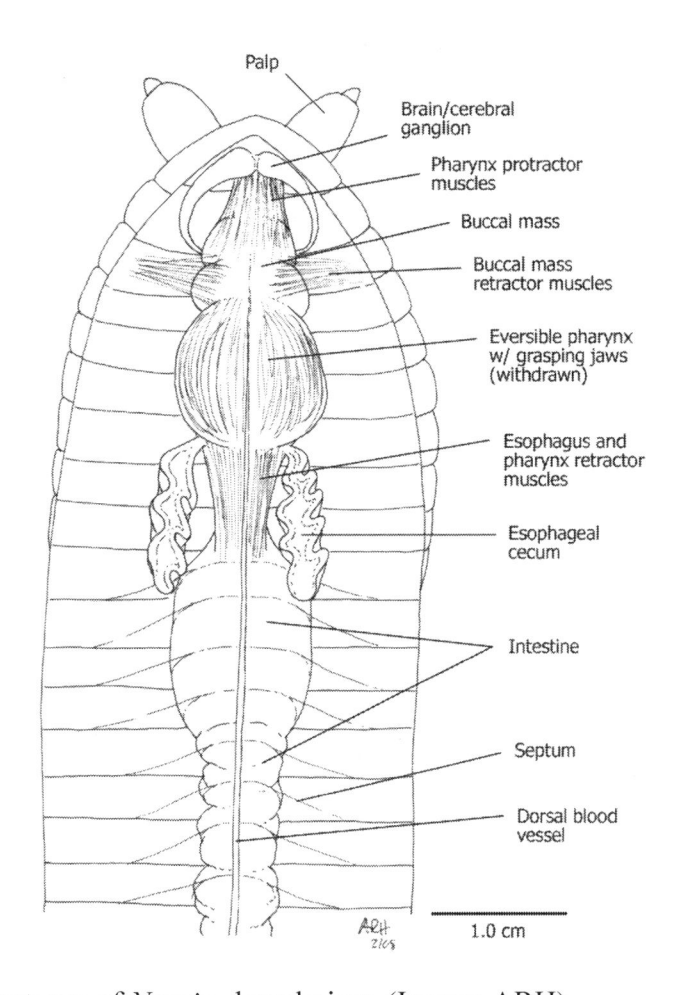

Figure 8.4. Internal anatomy of *Nereis*, dorsal view. (Image: ARH)

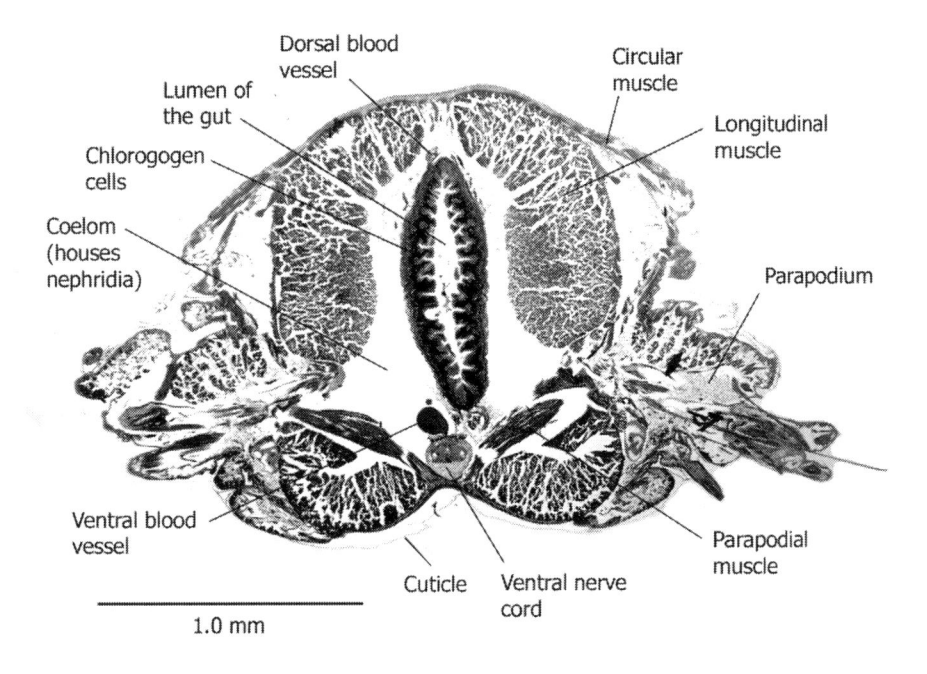

Figure 8.5. Cross-section through the intestinal region of *Nereis*. (Image: ARH)

Subclass Sedentaria, Family Clitellata – oligochaete worms and leeches

This exercise highlights two representatives of this group: an earthworm and a leech. Earthworms play an important role in nutrient cycling as they ingest soil, digest organic material from it, and deposit nutrient-rich feces. Their burrows aerate the soil and increase the availability of oxygen to plant roots.

Tasks – Earthworm

1) Obtain a live earthworm and observe its behavior. Rinse your specimen and place it on a moist paper towel so you can observe the contraction and elongation of individual body segments as it moves. Describe what you see.
2) Devise a simple experiment to see if earthworms prefer light or shade. Be sure to record your methods, results and conclusions in your lab notebook.
3) Discover the location of chaetae on the earthworm *Lumbricus* by gently running your fingers along the length of the worm from anterior to posterior and then from posterior to anterior. Repeat this for dorsal and ventral surfaces of the worm. Record your observations. Where are the chaetae located?
4) Place a live specimen on a dry paper towel and listen carefully as the worm moves. Listen for the scratching sound of chaetae on the paper towel. Do not leave the worm on a dry paper towel long because it will dry out quickly.
5) Obtain a preserved specimen, place it in a wax-bottom dissection tray and add enough water to cover it. Use a magnifying lens or dissection scope and observe its external anatomy. DRAW what you see. Use **Fig. 8.6** to help you identify visible structures. Be sure that structures on your drawing correspond to the correct segment numbers from the prostomium.

Figure 8.6. External anatomy of the anterior portion of the earthworm *Lumbricus*, lateral view. Segments 10 and 15 are indicated for reference. (Image: ARH)

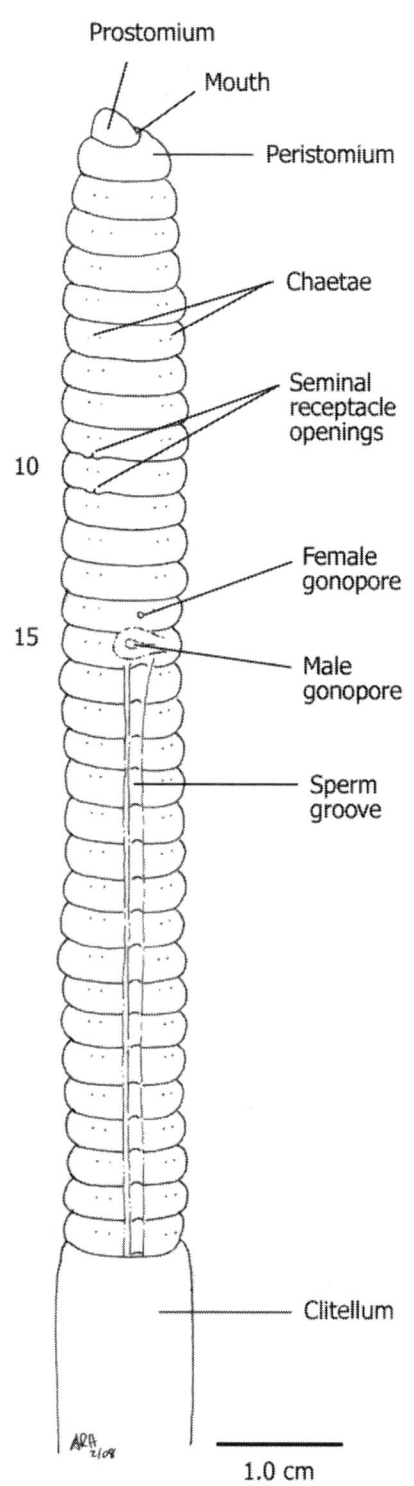

6) **Internal anatomy – *Lumbricus*.** Follow the same procedure for dissecting your earthworm as you did for *Nereis*. Pin your earthworm to the dissection tray, dorsal side up. You can tell which surface is dorsal and which is ventral by running your fingers lightly from the posterior end of the worm to toward the head. The surface that bears chaetae is ventral. DRAW what you see and refer to **Fig. 8.7** to help you identify anatomy of your specimen. You will probably not see everything indicated on **Fig. 8.7** because some structures are visible only when an individual is mature and ready to mate.

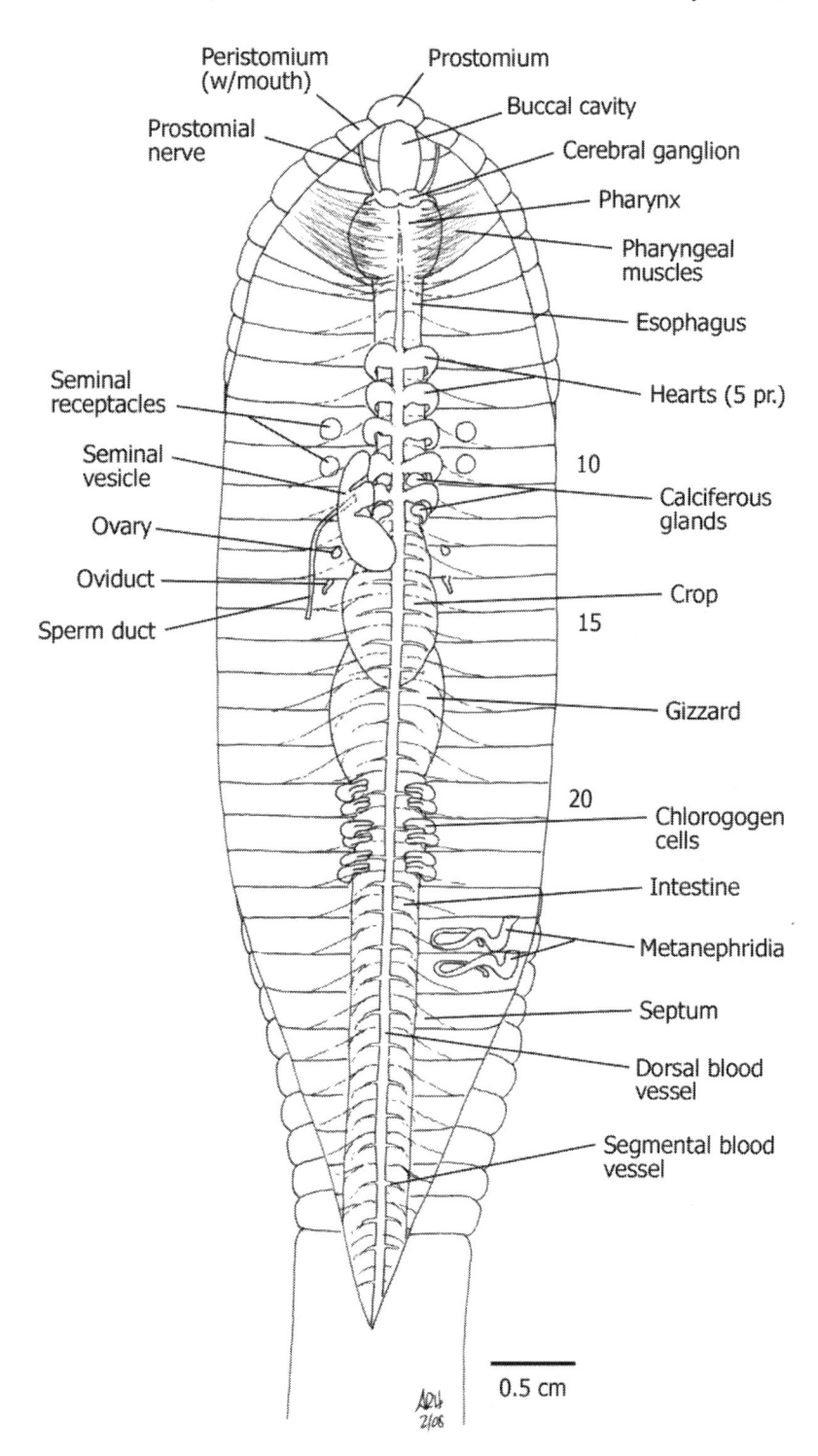

Figure 8.7. Internal anatomy of the earthworm *Lumbricus*, dorsal view. (Image: ARH)

7) Obtain a prepared cross-section slide of *Lumbricus* and observe it under a compound microscope. DRAW what you see and use **Fig. 8.8** to help you identify visible structures.

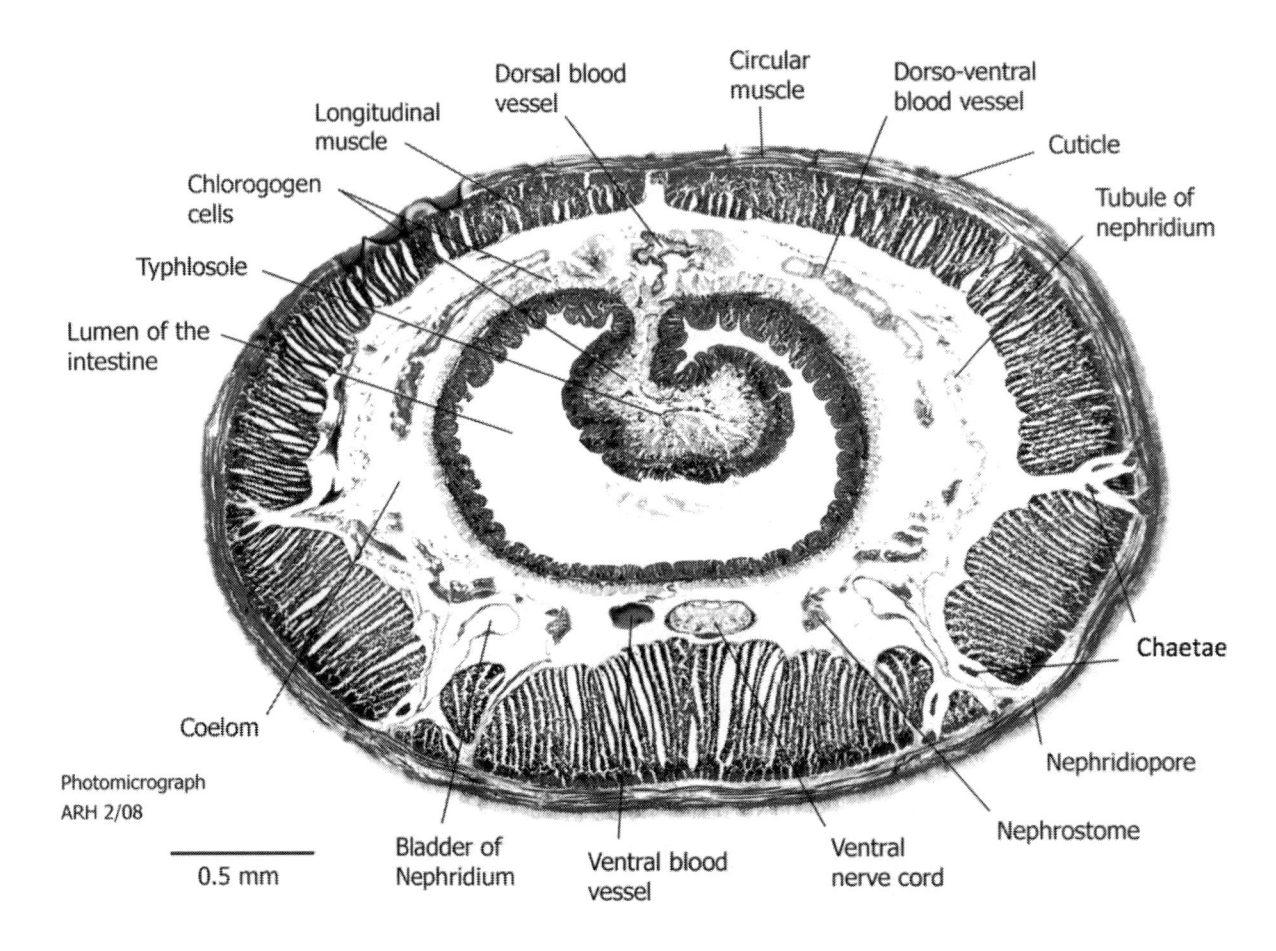

Figure 8.8. Cross-section of the earthworm *Lumbricus* through the intestine. (Image: ARH)

Tasks - Leech

1) External anatomy of a preserved leech. Use a magnifying lens or dissection scope to examine your specimen. The body wall appears to be divided into many segments, these are external annuli and there are usually over 90 of them. External annuli do not correspond to internal body segmentation. Leeches always have 33 internal body segments. DRAW the external anatomy of your leech and refer to **Fig. 8.9** to help you identify what you see.
2) Body wall organization. Use a dissection scope to examine the orientation of fibers in the body wall of the leech. Look at both the dorsal and ventral body walls as you do this. DRAW the orientation of the fibers in the body wall and then hypothesize about the benefits of this body wall anatomy.
3) You are not required to dissect the leech but if you have extra time you can give it a shot. Leeches are notoriously difficult to dissect because they have a reduced coelom and massive amounts of muscle in the body. Follow the procedure to open up the body of the leech as you did for *Nereis* and *Lumbricus*. You will need to refer to figures in your

textbook or material online to help you identify what you see. DRAW what you see and identify everything you can.

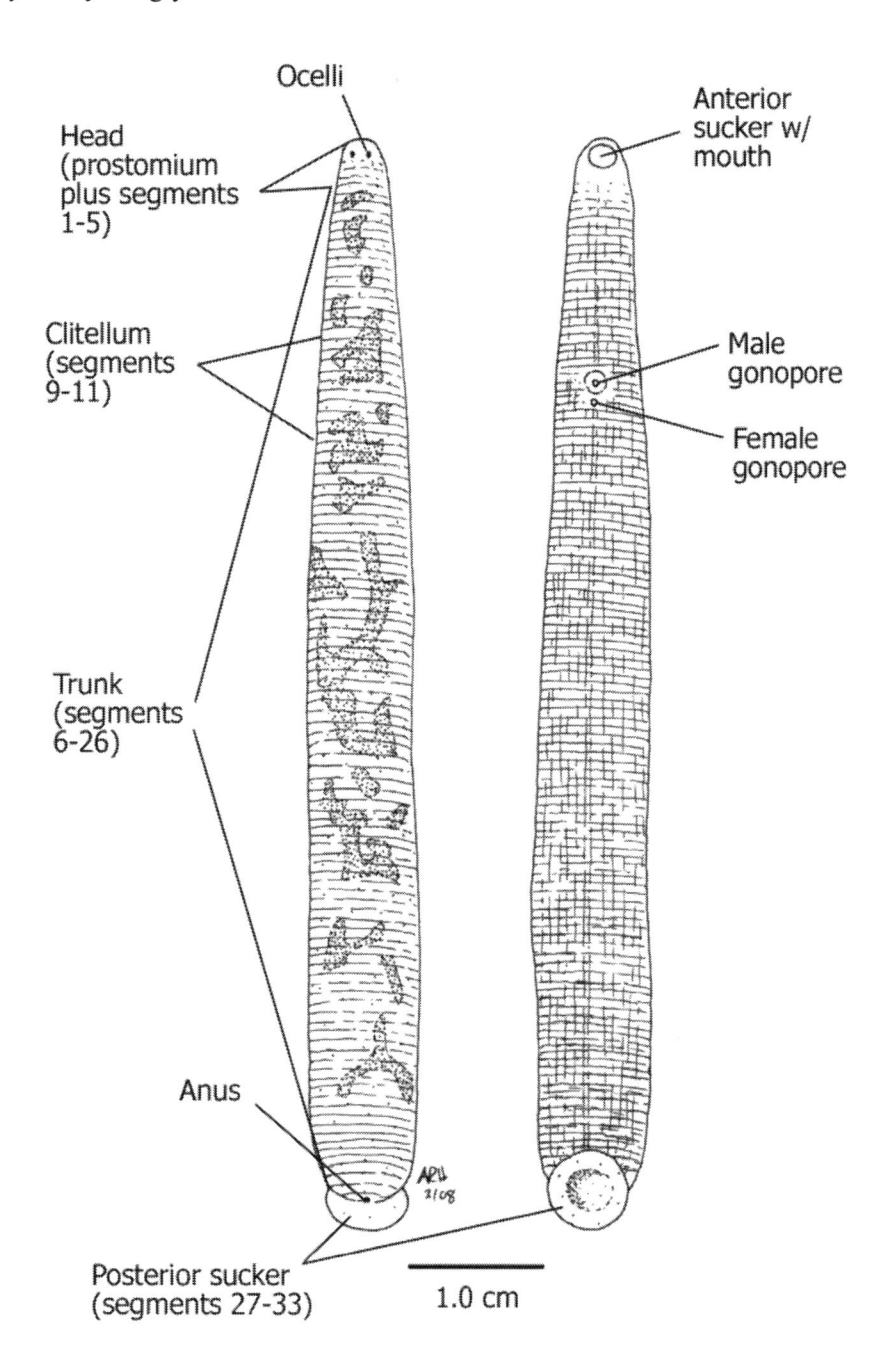

Figure 8.9. External anatomy of the leech *Hirudo*, dorsal view **left** and ventral view **right**. (Image: ARH)

Group questions

1) How is the anatomy of a leech adaptive to its ectoparasitic lifestyle?
2) Leeches have no chaetae, why is this advantageous to them?

3) Most of you have seen earthworms in a biology lab before but how has today's exercise expanded or changed your view of annelids?
4) What aspects of the body plan of annelids make them so successful ecologically?

Phylum Annelida – Glossary

Aciculum – stiff rod-like structure that provides support to parapodia, parapodial muscles attach to these to move parapodia

Anterior sucker (leeches) – attachment and feeding organ of leeches, surrounds the mouth

Bladder of nephridium – portion of a nephridium where urine can be stored temporarily

Buccal mass – anterior-most portion of the digestive tract

Buccal mass retractor muscles – pull the buccal mass and pharynx back into the body

Calciferous gland – organ that releases excess $CaCO_3$ into the gut

Chaetae – epidermal bristles made of sclerotized (hardened) chitin

Chlorogogen cells – a mass of cells lining the digestive tract that functions like a liver, storing and releasing nutrients as needed

Circular muscles – muscles just beneath the epidermis that run around the girth of an animal, these oppose the longitudinal muscles in animals that have a hydrostatic skeleton

Cirrus, dorsal and ventral (of parapodia) – extensions of parapodia that are sensory organs

Clitellum – region of the body that secretes the cocoon

Closed circulatory system – circulatory system where blood is always in vessels

Coelomate – a body that contains one or more fluid-filled cavities that are lined by mesodermal peritoneum

Contractile vessel – tube or vessel that is surrounded by muscles that contract and push fluid through it, sipunculans use these as compensation sacs - a structure where they can temporarily store excess fluid when the introvert is inverted, earthworms have contractile dorsal and ventral blood vessels

Crop – thin-walled organ where ingested material can be stored temporarily

Cuticle – thin non-cellular protective layer secreted by the epidermis

Esophageal cecum – pouches of the digestive tract

Eversible pharynx – portion of the digestive tract of annelids (including the jaws) that can be extended beyond the mouth

Gizzard – muscular organ that can grind up or mix ingested material

Hearts (earthworms) – five pairs of enlarged blood vessels that connect the ventral and dorsal blood vessels, they are contractile and help move blood through the body

Infauna – animals that live below the sediment surface

Introvert – eversible anterior organ used by sipunculans for feeding

Introvert retractor muscles – muscles that pull the sipunculan introvert into the coelomic space

Longitudinal muscles - muscles usually just beneath the epidermis and circular muscle layer that run along the long axis of an animal, these oppose the circular muscles in animals that have a hydrostatic skeleton

Metamerism – repeated structures in the body, e.g., segments in annelids

Metanephridia – excretory organ found in annelids that have a cilia-lined entrance, a tube where modification of the urine takes place and a bladder where urine is stored until it is released

Nephridiopore – opening through which urine is released

Nephrostome – opening into the metanephridium through which coelomic fluid is pulled

Neuropodium – large ventral lobe of a parapodium

Notopodium – large dorsal lobe of a parapodium

Nuchal organ – ciliated pit believed to be chemosensory

Palp – sensory organ of the head on some annelid worms

Parapodial muscle – a muscle that moves a parapodium

Parapodium – a fleshly muscular extension of the body wall that is used for walking, swimming or moving through a burrow

Peristomium – second segment of the annelid body, has a pair of coelomic spaces and houses the mouth and cerebral ganglia

Pharyngeal protractor muscles – muscles that extend the pharynx anteriorly, sometimes out through the mouth

Posterior sucker (leeches) – attachment and locomotory organ of leeches, the anus is located on the dorsal surface of this organ, not in it

Prostomial nerve – nerve that extends from the cephalic ganglion into the prostomium

Prostomium – first segment of the annelid body, bears sensory tentacles in some species and always contains only one coelomic space

Radial fixing muscles – muscles that attach the intestine to the body wall in sipunculans

Segmentation – body divided into discrete, repeated compartmentalized spaces

Seminal receptacle – female organ that receives sperm during mating

Seminal vesicle – male organ that stores sperm until it is used during mating

Septum – wall of peritoneal tissue that separates neighboring segments in the annelid body

Sperm groove – sperm move along this longitudinal external depression in the body wall between the male gonopore and clitellum where the cocoon is formed and eggs are fertilized

Spindle muscle – muscle in sipunculans inserted in the posterior body wall that runs up through the center of the coiled intestine and compensation sacs, this muscle prevents the intestine from becoming tangled or twisted

Tentacles – unsegmented tactile and chemosensory organs

Tentacular cirri – tactile and chemosensory organs

Triploblastic – body that produces embryonic ectoderm, endoderm and mesoderm

Tubule of metanephridium – portion of a metanephridium where the urine is modified by secretion and reabsorption of materials

Typhlosole – infolding of the intestinal wall that is attached dorsally along its length and produces a U-shaped lumen in cross-section, this infolding of tissue houses chlorogogen cells

Chapter 9: Phylum Brachiopoda and Phylum Nematoda

This may seem like an unusual pairing of taxa to cover in one lab exercise but time constraints of most academic calendars make a chapter like this one inevitable. Even so, this pairing may not be that surprising considering what you have already done and what is coming up next. Phylum Brachiopoda is a member of Clade Spiralia, as are flatworms, molluscs and annelids. Phylum Nematoda is a member of Clade Ecdysozoa along with arthropods and their other relations.

Phylum Brachiopoda belongs to Clade Lophophorata, a clade within Clade Spiralia. Lophophorates are animals that make a living by using a ciliated crown of tentacles called a lophophore for suspension feeding and gas exchange. Clade Lophophorata includes the following phyla: Phoronida (phoronids), Bryozoa (ectoprocts or bryozoans) and Brachiopoda (lamp shells). You will examine brachiopods in this exercise.

The other taxon in this chapter is Phylum Nematoda. These are the roundworms and are the first Phylum of Clade Ecdysozoa covered in this laboratory manual. Ecdysozoans lack locomotory cilia and they secrete a tough chitinous cuticle or exoskeleton that is molted a specific process called ecdysis. Clade Ecdysozoa includes three smaller clades. Phylum Nematoda (roundworms) and Phylum Nematomorpha (horsehair or Gordian worms) make up Clade Nematoida. The other two clades are Clade Scalidomorpha and Clade Panarthropoda. Clade Scalidomorpha includes Phylum Priapulida (priapulid or penis worms), Phylum Loricifera (loriciferans – no common name), and Phylum Kinorhyncha (mud dragons – you've got to love that there are animals out there called mud dragons!). Clade Panarthropoda includes the Phylum Tardigrada (water bears – perhaps the cutest animals ever), Phylum Onychophora (velvet worms) and Phylum Arthropoda (arthropods).

Clade Lophophorata "tuft bearing" - Phylum Brachiopoda "gill foot" - brachiopods

Brachiopods belong to Clade Lophophorata and include living species and many important index fossils. Index fossils are species that had a large geographic range, produced many fossils and were present in the fossil record for a limited time before they were replaced by other species.

Brachiopods were once the dominant shelled suspension-feeding taxon in the sea. Bivalve molluscs have since largely displaced them but brachiopods still survive in deep sea and other refuge habitats. Characteristics of Phylum Brachiopoda are listed in Table 9.1.

Table 9.1. Characteristics of Phylum Brachiopoda (after Brusca, *et al.*, 2016).

- Lophophore
- Dorsal and ventral shells
- Mantle lobes and mantle cavity
- Pedicle
- U-shaped gut
- Benthic, solitary animals

Brachiopods bear a crown of ciliated tentacles called a lophophore, the organ used for feeding gas exchange. Lophophore size and shape varies from species to species. You are introduced to two forms of brachiopods in this exercise. The first one is the infaunal inarticulate brachiopod *Lingula*. This is one of the oldest known living species of animals. Fossils of *Lingula* from the Cambrian Era over 500 million years ago are virtually indistinguishable from living specimens. The other form is an articulate brachiopod, either *Terebratalia* or *Terebratalia*. By the way, all brachiopods have two shells that are dorsal and ventral instead of being left-hand and right-hand shells as in bivalves.

Phylum Brachiopoda – Taxonomy

The taxonomy of brachiopods has been a focus of zoologists and paleontologists for many years. This is because brachiopods are such an important taxon in interpreting the biological and geological history of earth. As mentioned above many extinct species are important index fossils because they left such a rich fossil history. The sizes, shapes and sculpturing of their shells that fossilize well are important characteristics. Brachiopods are currently assigned to two subphyla.

Subphylum Craniiformea "helmet form" (covered in this exercise) – inarticulate brachiopods. These animals lack tooth-and-socket hinges and use muscles to attach shells together. These shells are made of $CaCO_3$ and proteins and the ventral valve is cemented directly to a hard substrate. Most lack a pedicle but *Lingula* is an exception because it produces a pedicel and is infaunal. There are 20 extant species in this taxon.

Subphylum Rhynchonelliformea "all nose form" (covered in this exercise) – articulate brachiopods. These animals have a tooth-and-socket hinge to attach shells together, their shells made of proteins and $CaCO_3$ and they use a pedicel to attach themselves to hard substrates. There are 350 extant species in this taxon.

Tasks – *Lingula* (inarticulate brachiopod)

1) Use a magnifying glass or dissection scope to study the shell and external anatomy of *Lingula*. Detach the pedicle from the shell and make a cross-sectional cut through it and examine its anatomy. DRAW your specimen's external anatomy. Refer to **Fig. 9.1** to identify what you see.

Figure 9.1. External anatomy of *Lingula*. (Image: ARH)

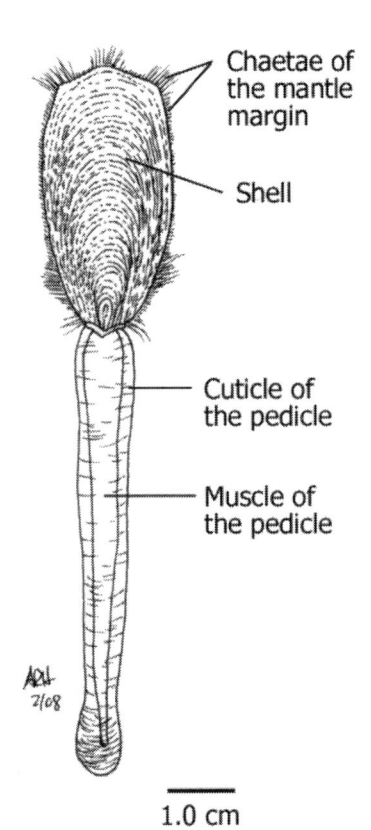

2) Anatomy of the mantle cavity. Unlike bivalves that have a pair of right-hand and left-hand shells, brachiopod shells are dorsal and ventral in orientation and you can't readily tell which is shell is dorsal and which is ventral in *Lingula* until you separate the shells. Use a probe tip to separate the mantle and anterior adductor muscles from the inside of one shell. Gently pull the shells apart or use forceps to crack and peel away small pieces of one shell until the structures inside the mantle cavity are exposed. The lophophore is always attached to the dorsal valve of the shell. Remove the ventral shell, exposing the lophophore, muscles and other soft tissues. Doing this exposes the mantle and mantle cavity. Branches of the coelom extend to near the margin of the mantle. DRAW what you see and use **Fig. 9.2** (left image) to help you identify anatomical structures.

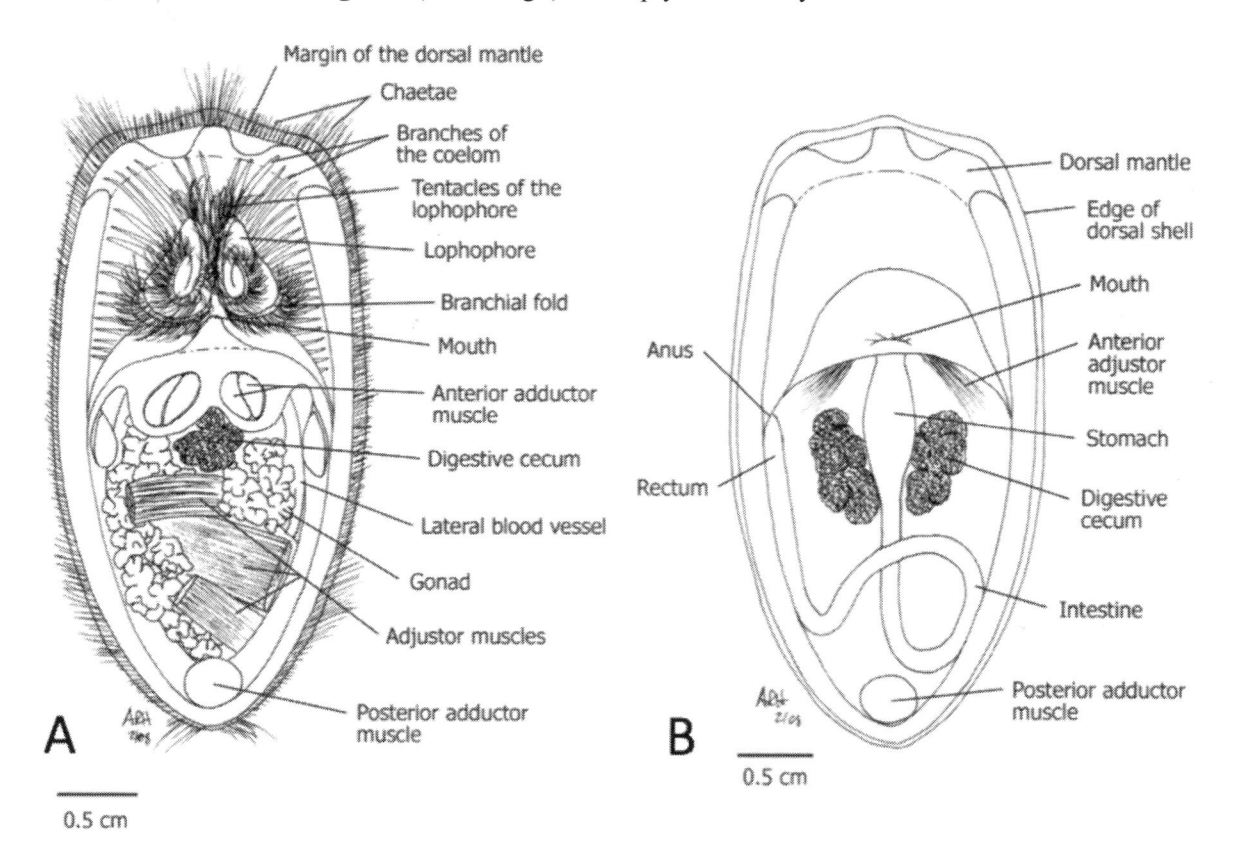

Figure 9.2. Internal anatomy of *Lingula*: **A)** lophophore and superficial soft tissues, **B)** digestive tract and deep tissues. (Images: ARH)

3) Superficial soft tissues. The posterior portion of the shell near the attachment point with the pedicle houses the viscera. You should be able to locate the greenish digestive cecum, yellowish gonads and orange-red lateral blood vessels. Locate the anterior and posterior adductor muscles. The anterior adductor muscles are divided into two parts: catch muscle fibers that contract quickly and fibers that contact slowly. The slowly contracting muscle fibers can remain contracted for prolonged periods of time and are used to keep the shells from being prying apart after they close. Look for three sheet-like oblique adjustor muscles. These hold the shell valves in correct position relative to each other since *Lingula*. Remember that this is an inarticulate brachiopod and its shells do not have

hinge teeth. DRAW what you see and use **Fig. 9.2A** to help you identify structures of the visceral mass.

4) Digestive tract. Remove the digestive cecum, gonads, lateral vessels, anterior adductor muscles and oblique adjustor muscles. Doing this will uncover structures of the digestive tract and the anterior adjustor muscles. DRAW the digestive tract and other exposed soft tissues. Use **Fig. 9.2B** to help you to identify what you see.

Tasks – *Terebratalia* or *Terebratulina* (articulate brachiopods)

1) Examine the external anatomy of your specimen. DRAW the shell of your specimen and look **Fig. 9.3A** to help you identify what you see.
2) Insert a scalpel blade or probe tip between the dorsal and ventral shell valves of your specimen and pry them open far enough to use your fingers to gently pull them apart. Pull on the dorsal shell valve until you feel the hinge start to give way, and then carefully separate the shell valves from each other. This exposes the structures in the mantle cavity. Immerse your specimen. Take time to observe the structures associated with the ventral valve as well as the lophophore that is attached to the dorsal valve as well as other structures associated with the dorsal valve. DRAW what you see (refer to **Fig. 9.3B-C**).
3) Digestive tract. Remove the lophophore and adductor muscles from your specimen. This exposes structures of the digestive tract. DRAW what you see and use **Fig. 9.3D** to help you identify these structures.

Figure 9.3. Anatomy of the articulate brachiopod *Terebratalia*: **A)** shell anatomy, **B)** mantle and branches of the coelom, **C)** lophophore and superficial soft tissues, **D)** digestive tract and deep soft tissues,. (Image: ARH)

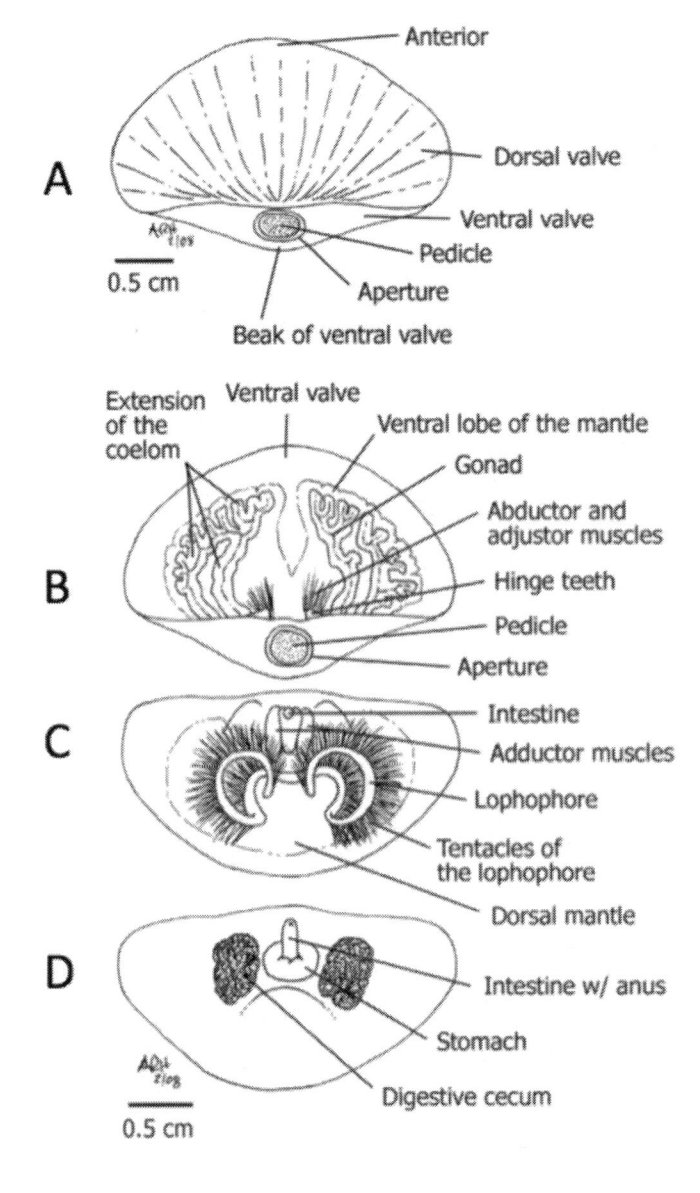

4) DRAW a fossil brachiopod and compare it to a preserved specimen. WRITE your observations in your lab manual.

Clade Ecdysozoa "slipping out animals" - Phylum Nematoda "thread-like" - roundworms

Roundworms may be the most numerically abundant kind of animal on the planet. We know quite a lot about medically important roundworms that are parasites but there are vast numbers of non-parasitic species living in the soil and other habits about which we know next to nothing. This makes estimating their diversity difficult. Most nematodes live in the soil or other habitats where they are benign to human health.

Nematodes undergo ecdysis only a few times during their larval development. One of the clues that roundworms belong in the same clade as arthropods is that these groups both carry out ecdysis using the same hormonal control system. Follow-up molecular phylogenetic analysis confirmed that nematodes are members of Clade Ecdysozoa.

The taxonomy of nematodes is quite complex and is not covered in this manual.

Nematodes have a relatively simple anatomy. They have a tough outer cuticle that is made of β-chitin, four blocks of longitudinal muscles in the body wall and a blastocoelom (pseudocoelom) that may be quite spacious in some species and highly reduced in others, especially in tiny species. Many nematodes also exhibit a trait called eutely, i.e., they produce a predetermined number of cells for an organ or for the entire body. This trait makes the roundworm *Caenorhabditis elegans* a useful species for doing research in developmental biology. Characteristics of nematodes are listed in **Table 9.2**.

Table 9.2. Characteristics of Phylum Nematoda (after Brusca, *et al.*, 2016).

- Triploblastic, unsegmented worms
- Blastocoelomate (pseudocoelomate) body plan
- Body is round in cross section
- Ecdysis of cuticle only a few times during juvenile stages
- Amphids (cephalic sense organs) and phasmids (caudal sense organs)
- Body wall with only longitudinal muscles

Tasks – *Tubatrix aceti*

1) Roundworm behavior. Live specimens of the vinegar eel *Tubatrix aceti* provide good insights into how roundworms move. Place some live vinegar eels in a small glass bowl with enough liquid to allow them to move freely or make a wet mount side observe them using either a dissection or compound microscope. Keep in mind that roundworms have only longitudinal muscles in their body walls. WRITE your observations of how vinegar eels move and compare their movement to that of the earthworms you observed in another exercise.

Tasks – *Ascaris lumbricoides*

1) External anatomy of *Ascaris*: Work with a lab partner during this portion of the lab and be sure to glove up because female *Ascaris lumbricoides* contain thousands of embryonated eggs that can still be viable even after being immersed for prolonged periods of time, even in toxic preservatives. *Ascaris lumbricoides* is the largest intestinal roundworm parasite of humans, reaching lengths of 30 cm. Obtain preserved specimens of a female and a male worm. Females are usually longer and larger than males and females are pointed at both ends. Males are smaller and the body forms a pronounced hook at the posterior end. Use a magnifying lens or dissection scope to locate the three lips at the anterior end of the body. *Ascaris* has one dorsal lip and two ventral-lateral lips. You can determine which the dorsal lip is by using a magnifying lens or dissection scope to find them and the lateral epidermal cords that run along the sides of the body of *Ascaris*. The lip that is medial relative to the lateral cords is the dorsal lip. Once you have identified the dorsal surface use insect pins to attach your worm dorsal surface up to a dissection tray. DRAW the external anatomy of your worms.

2) Internal anatomy of *Ascaris*. One of you should dissect a female worm and the other should dissect a male worm. Show each other what you see as you work through these dissections. Use a scalpel to open the body cavity by making a longitudinal incision through the dorsal body wall that runs the length of the body. The body wall is thin and the structures inside the body cavity are delicate so be careful as you cut. Fold the body wall back and use insect pins to attach the body wall to the bottom of your dissection tray. If you are dissecting a female worm continue your incision for the entire length of the body. If you are dissecting a male stop your incision short of the hooked posterior end. DRAW the internal anatomy of your specimen. You may need to gently tease apart some structures in order to see them and their spatial relationships to each other. Look at **Figs. 9.4** and **9.5A** to help you identify what you see.

3) Embryonated eggs. Make a wet-mount slide of contents of the uterus of a female worm. Infertile eggs have an outer covering that looks like it contains many small droplets. A fertile egg contains a single large cell with a visible nucleus. Some egg capsules may contain embryonic worms. DRAW and describe what you see.

4) Examine the male reproductive system. Cut the hooked end off of the male worm. Make a longitudinal cut directly through that structure as shown in **Fig. 9.5B**. Use a dissection scope to see if you can identify structures of the male cloacal region – these are sometimes difficult to see and are visible only if you make a cut that runs right through them.

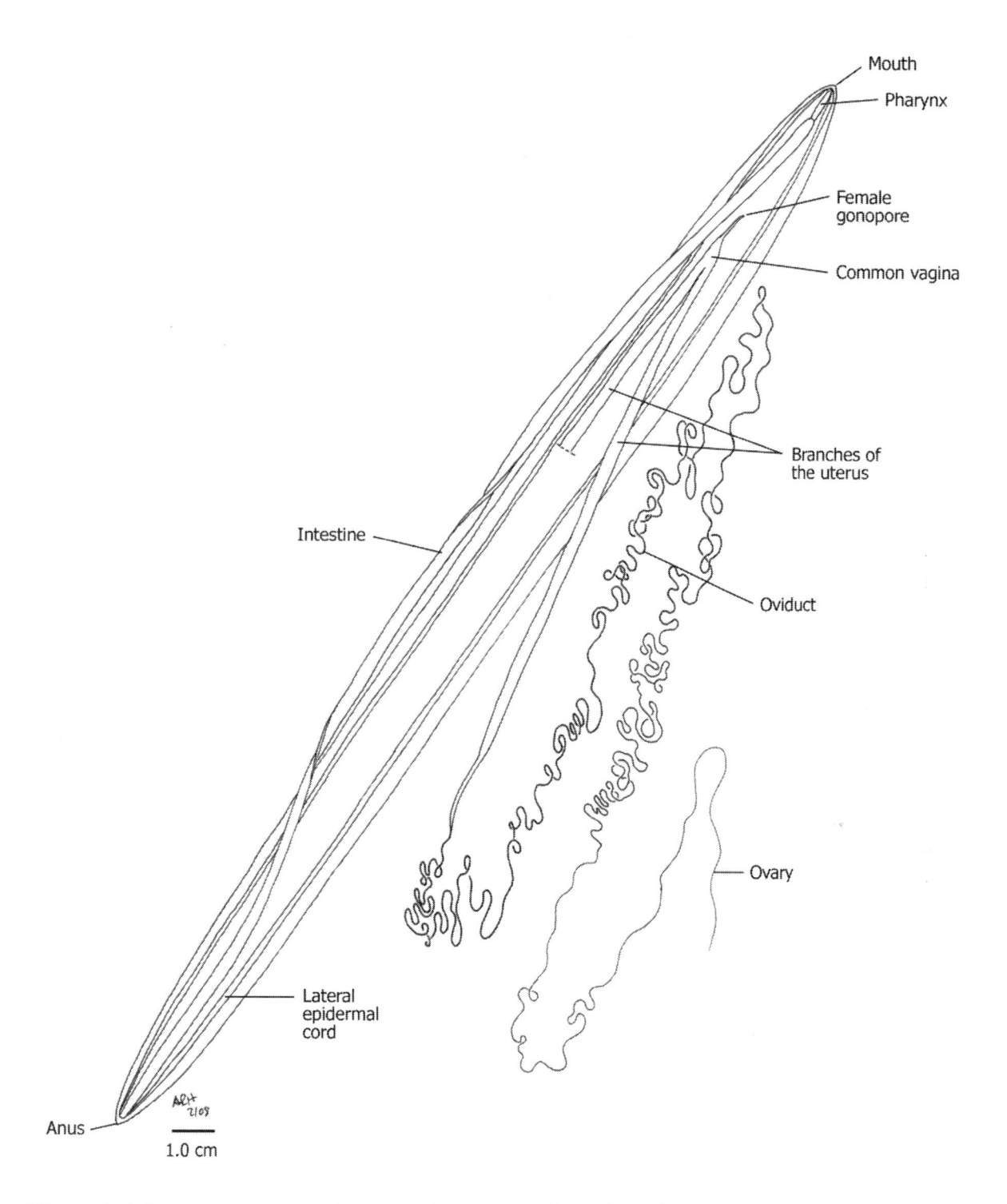

Figure 9.4. Internal anatomy of female *Ascaris*; one branch of the uterus and associated organs are not shown to simplify the image. (Image: ARH)

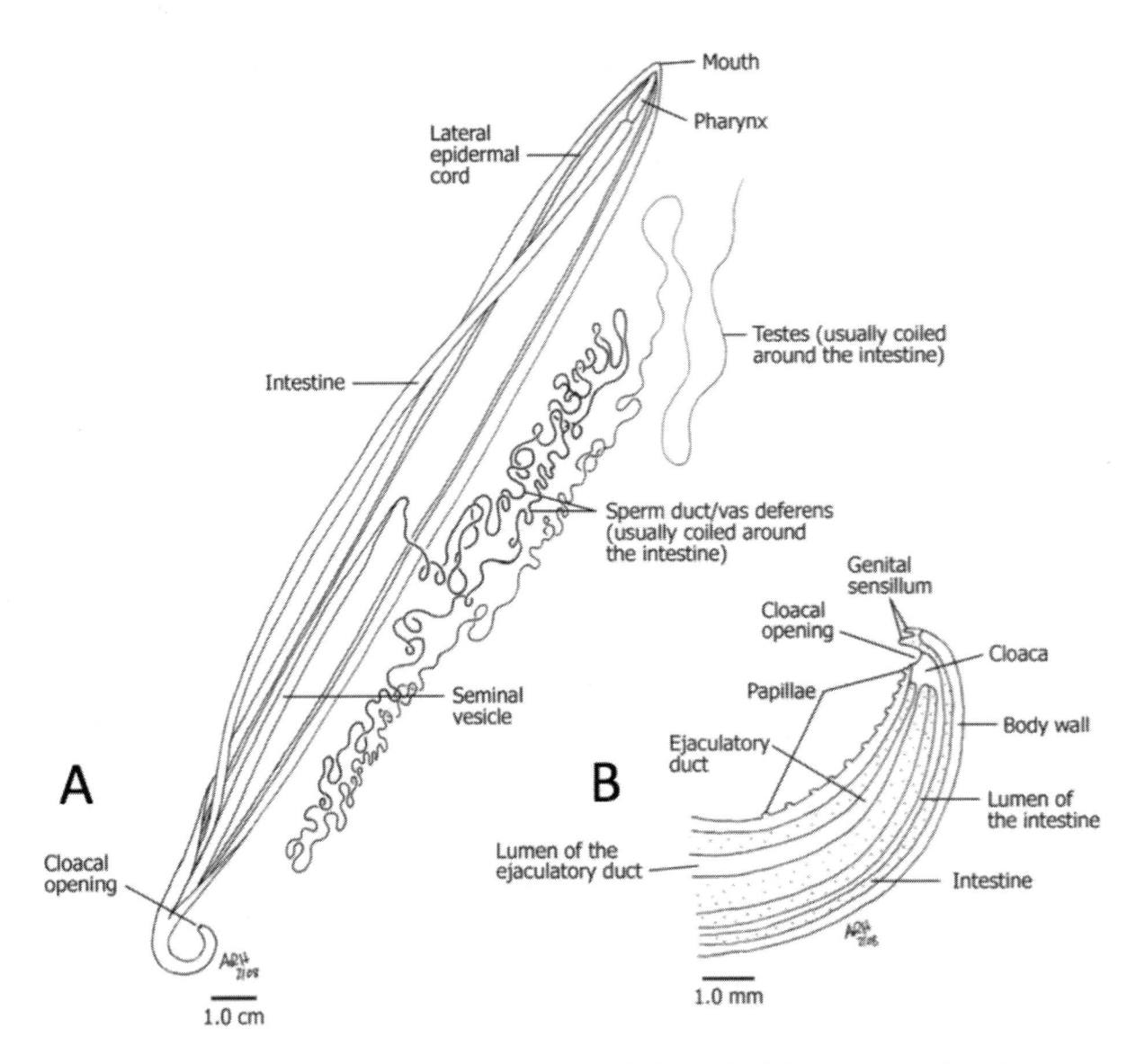

Figure 9.5. A) Internal anatomy of male *Ascaris*, and **B)** detail of the cloacal region. (Image: ARH)

5) Use a compound scope to examine prepared cross-section slides of *Ascaris*. Study cross-section slides of male and female *Ascaris* made through the intestine as well as a cross-section slide through the pharynx of *Ascaris*. DRAW what you see and use **Figs. 9.6-9.8** to help you identify the anatomy on these slides.

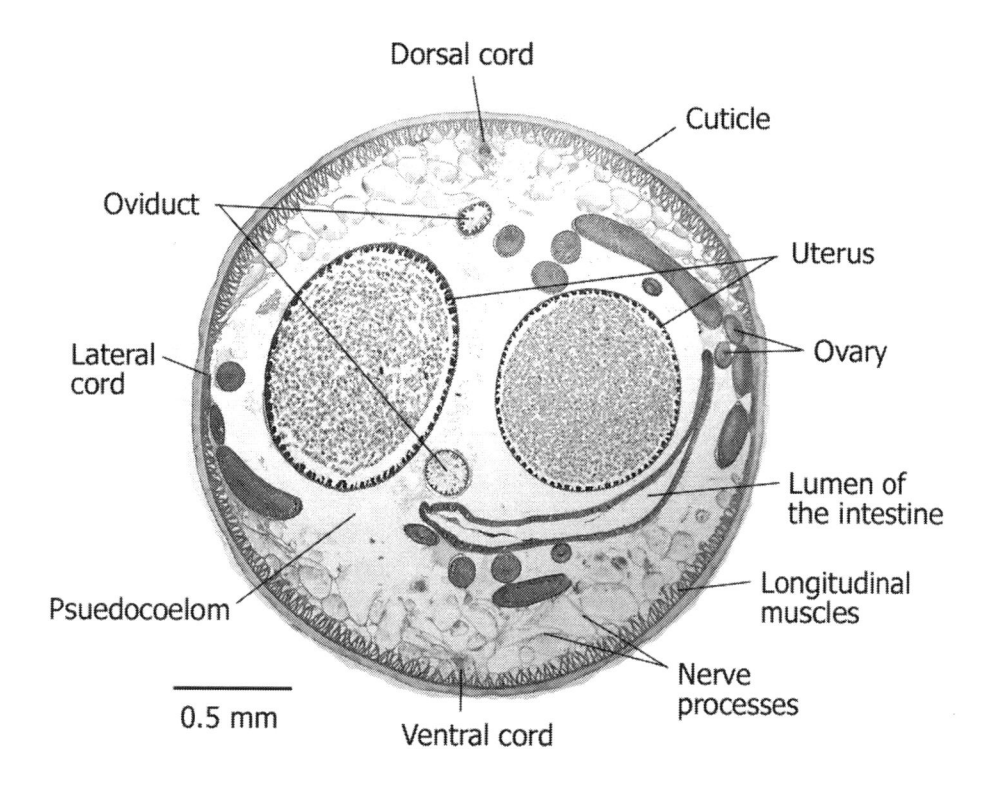

Figure 9.6. Cross-section through the intestinal region of female *Ascaris*. (Image: ARH)

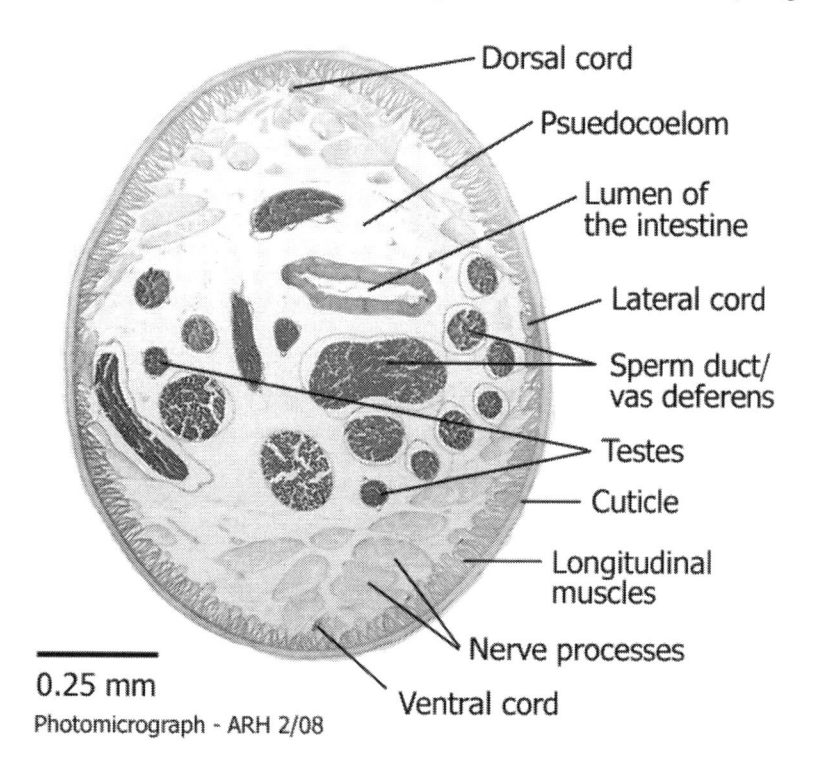

Figure 9.7. Cross-section through the intestinal region of male *Ascaris*. (Image: ARH)

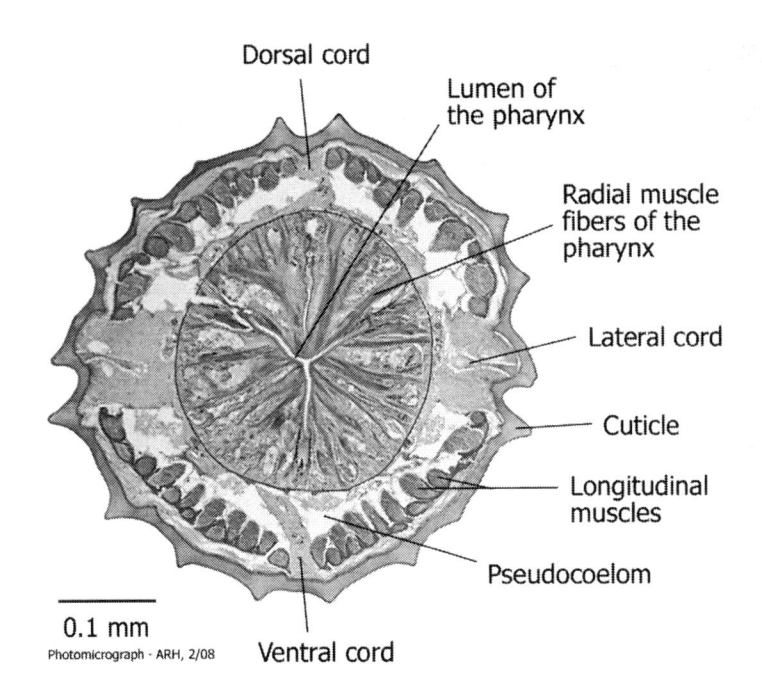

Figure 9.8. Cross-section through the pharynx of *Ascaris*. (Image: ARH)

6) Review the life cycle of *Ascaris*. See **Fig. 9.9**.

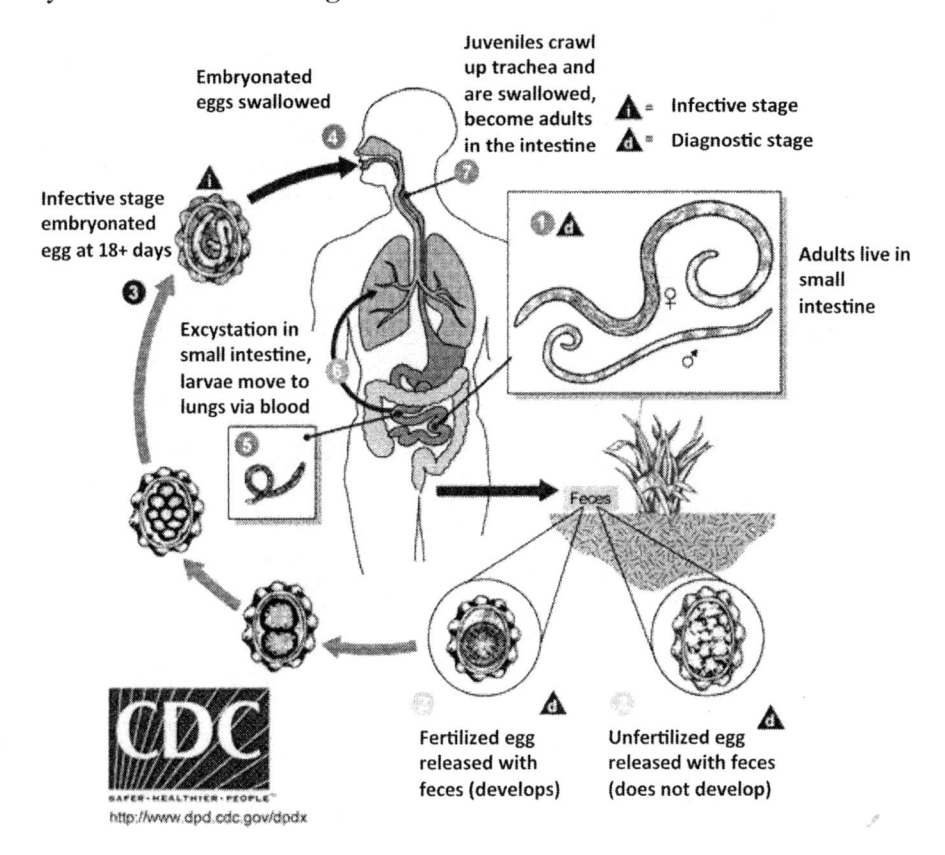

Figure 9.9. Life cycle of *Ascaris*. Adult worms live in the small intestine where they mate (1), eggs are released with feces and develop outside of the body (2-3), infective larvae are ingested (4) and excyst in the small intestine (5), they burrow through the intestinal mucosa and are carried by the blood to the lungs (6), they feed and grow in the lungs and when they approach maturity they crawl up the trachea to the pharynx where they are swallowed (7) and become adults in the small intestine. (Image: ARH modified CDC image, http://www.cdc.gov/dpdx/ascariasis/)

Tasks – Selected medically important nematodes

Look at the prepared slides and life cycles of any available medically important nematodes. A few representatives are included below:

1) *Trichinella spiralis*. This common human parasite causes a condition called trichinosis. Trichinosis occurs when a person eats raw or improperly cooked meat, usually pork that contains infective stage larvae encysted in the flesh. Disease symptoms include abdominal pain and gastrointestinal problems within the first few days with chronic symptoms including muscle pain, intestinal problems, general weakness, fatigue, headaches, etc. The latter symptoms are caused when next generation worms destroy host muscle tissue as they burrow in and encyst. This parasite is the reason that we are taught to cook pork thoroughly. The U.S. pork industry has nearly eradicated trichinosis in pork grown indoors but hogs grown outdoors often have this parasite so keep on cooking those pork products! **Figure 9.10** shows a cross section of infective *Trichinella worms* encysted in muscle tissue. **Fig. 9.11** shows the life cycle of *Trichinella spiralis*.

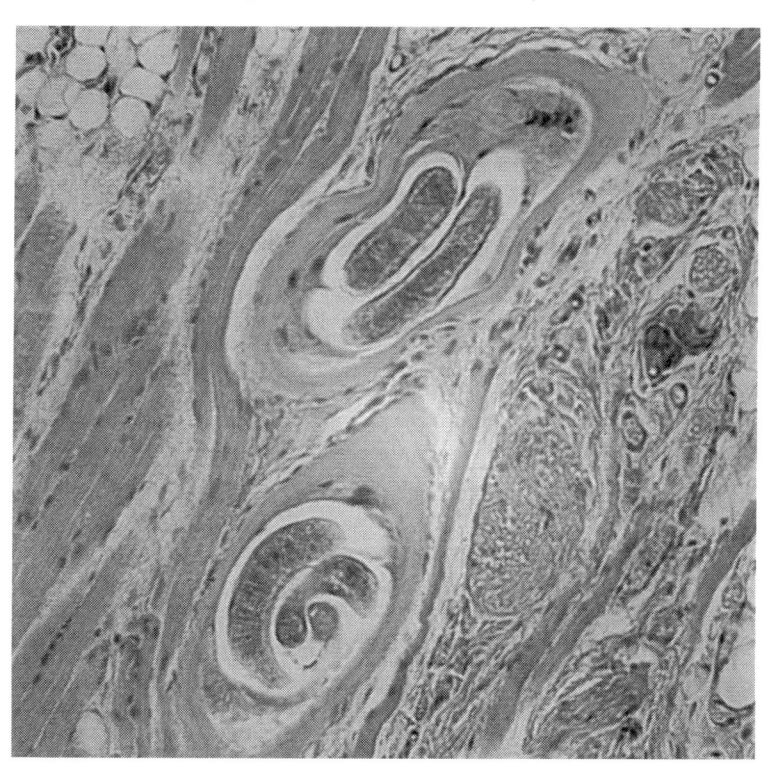

Figure 9.10. Cross-section of two encysted *Trichinella spiralis* larvae in muscle. (Image: ARH)

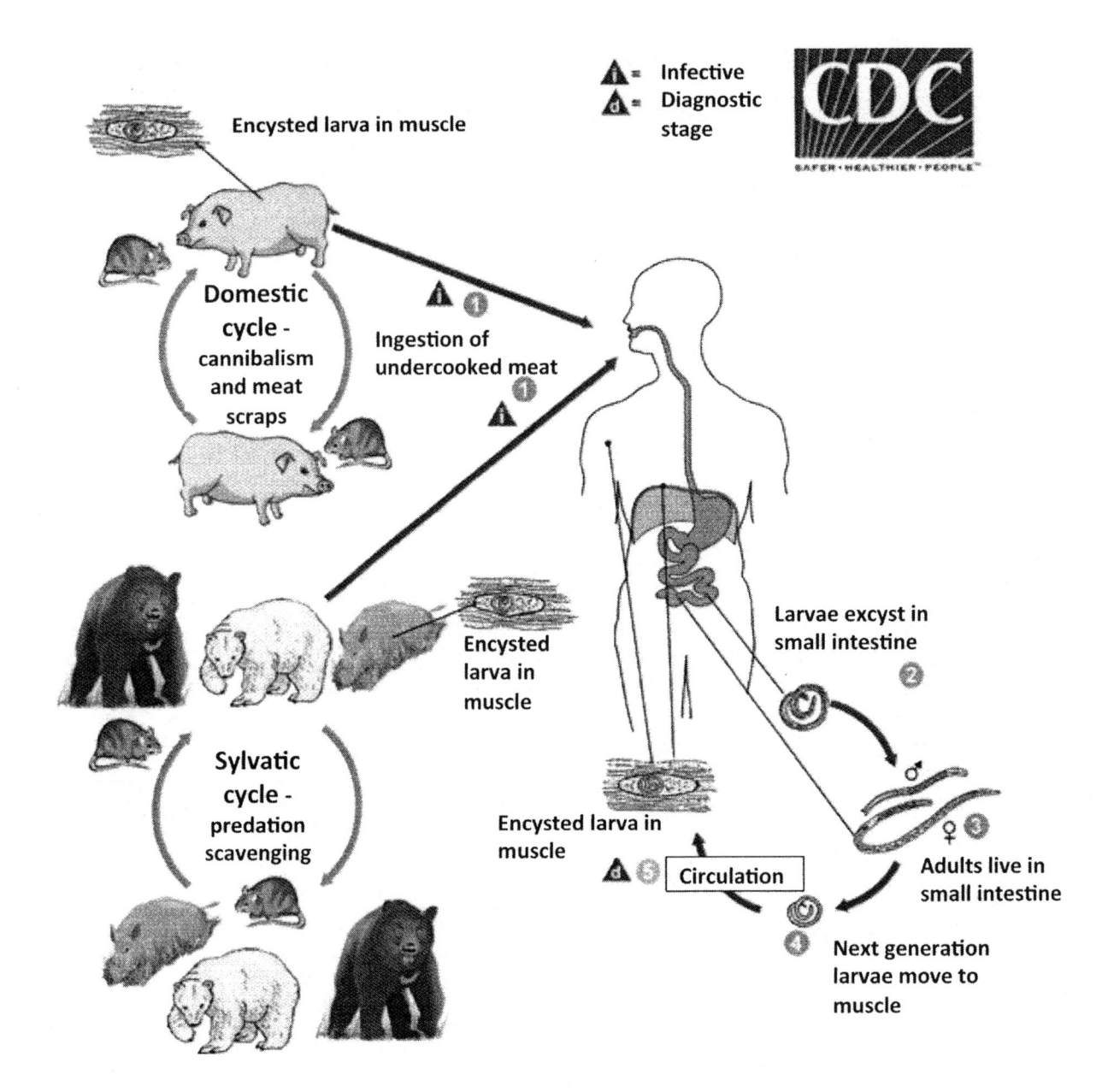

Figure 9.11. Life cycle of *Trichinella spiralis*. Infection occurs when humans ingest improperly cooked meat bearing encysted larvae (1). Larvae excyst, mature and mate in the small intestine (2-3). Next generation larvae burrow through the mucosa of the intestine (4) and into the bloodstream that carries them to skeletal muscle where they burrow in and encyst, destroying some muscle tissue as they do so (5). This life cycle is perpetuated among domestic and wild (sylvatic) animals through predator-prey or scavenging interactions in the wild and through cannibalism and ingestion of infected meat in domestic populations, mainly in populations of swine. (Image: CDC, http://www.cdc.gov/parasites/trichinellosis/biology.html)

2) *Wuchereria bancrofti.* This filarial roundworm is a species that causes elephantiasis. **Figure 9.12** shows the sheathed microfilarial stage in the blood. This is the stage that is taken up by mosquitoes when they feed on an infected host. **Figure 9.13** shows the life cycle of this species.

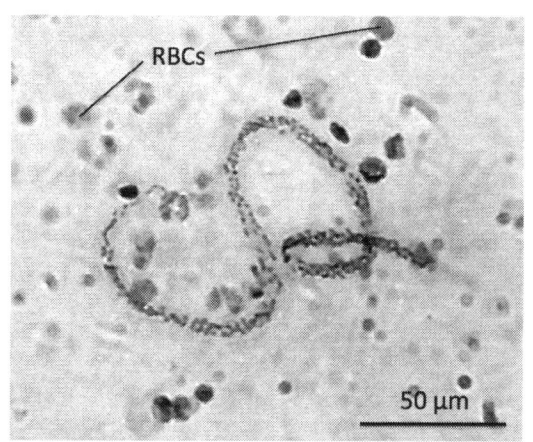

Figure 9.12. Blood smear preparation showing a *Wuchereria bancrofti* microfilaria stage larva. (Image: ARH)

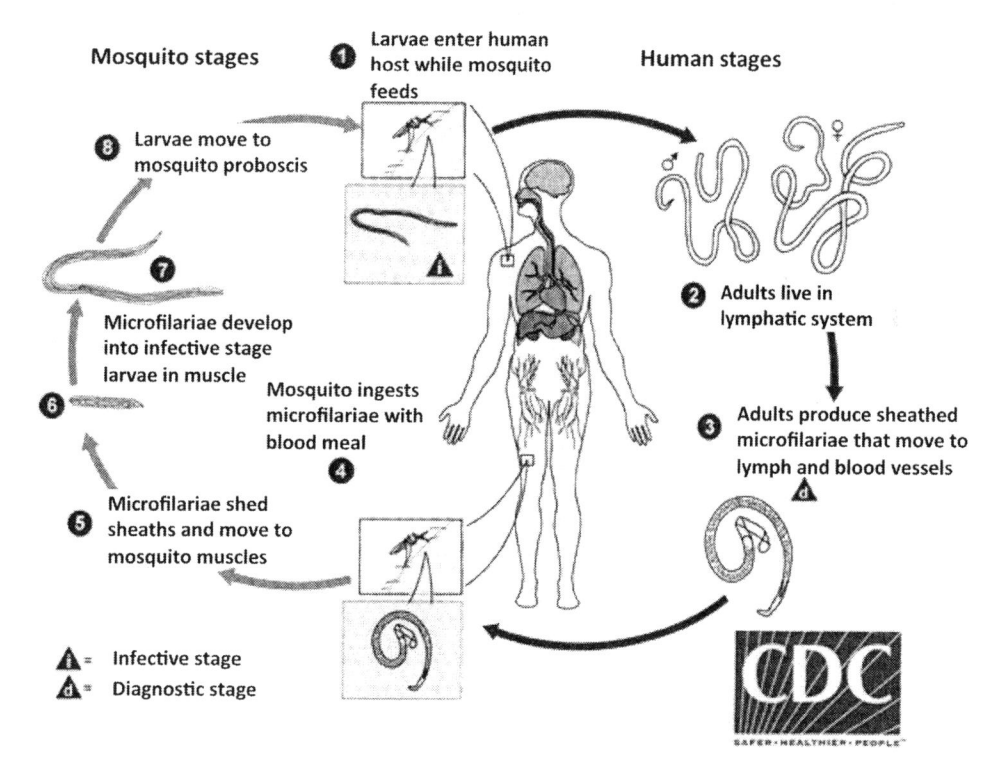

Figure 9.13. Life cycle of *Wuchereria bancrofti.* Infective larvae are transferred to a human when a mosquito feeds (1). Larvae make their way to the lymphatic system where adults take up residence in the lymph nodes (2), adults clog up the lymph nodes causing the symptoms of elephantiasis. Females can reach 8-10 cm in length and males are about half that size. Adults produce microfilariae that get into the blood stream where they are ingested by mosquitoes (3-4). Microfilariae develop into infective stage larvae in the gut of mosquitoes and then migrate to the proboscis of mosquitoes (5-8). According to the CDC elephantiasis can be treated with a drug that kills microfilariae and some adult worms. (Image: ARH modified CDC image, http://www.cdc.gov/parasites/lymphaticfilariasis/biology_w_bancrofti.html)

3) *Enterobius vermicularis.* Every prospective parent should know about this parasite. It is commonly known as pinworm and is the most common roundworm parasite in the USA. The pinworm lifecycle is shown in **Fig. 9.14**.

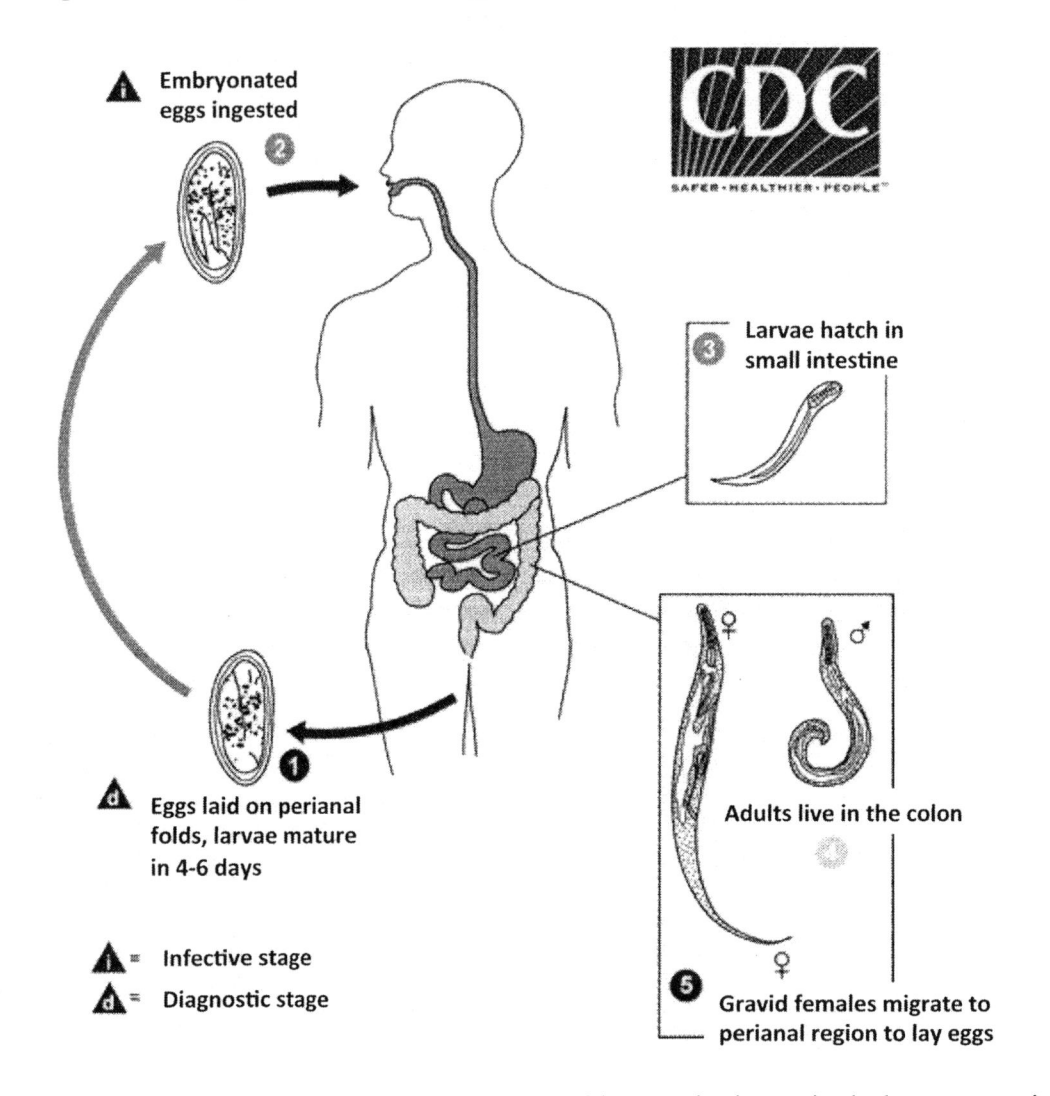

Figure 9.13. Life cycle of *Enterobius vermicularis*. This parasite is particularly common in children and among caregivers of children (especially those who bite their nails – think about it). Females emerge through the host's anus when the host is quiet, usually when the host is asleep. Females deposit embryonated eggs on the perianal region and retreat back into the rectum through the anus (1), these larvae become infective in 4-6 hours and are incidentally ingested (2), they hatch in the small intestine (3) and become adults and live in the colon and gravid females emerge to deposit eggs as frequently as nightly (4-5). Eggs can cause itching in the perianal region and this can lead to direct reinfection. Eggs can also get onto clothing, bedding, etc., and lead to indirect infection. Heavy infection can produce chronic lower abdominal pain. The easiest way to check to see if someone (e.g., a child) has this parasite is to wait a few hours after they have gone to sleep, slip down their underclothes and use a flashlight to see if there are any worms around the perianal region. They will appear as moving whitish threadlike objects up around 1.0 cm long. An anti-helminthic drug treats this condition effectively. (Image: ARH modified CDC image, http://www.cdc.gov/parasites/pinworm/biology.html)

4) *Dracunculus medinensis*. The Guinea fireworm is on the verge of being eradicated globally, mainly through efforts of the Carter Center – an initiative of former US President Jimmy Carter. If it is eradicated this will be only the second human disease to be completely eradicated. The first? Smallpox. The Guinea fireworm has a historic range covering parts of tropical Africa and Asia. Only a few cases per year are now being reported so eradication is within reach. You can learn more about the program and efforts to eradicate this roundworm parasite, a worm that has no treatment except slowly pulling the 1 meter long worm out of the body a few inches at a time by watching this video: https://www.youtube.com/watch?v=u4kQWvUv_Ns. **Figure 9.14** shows the life cycle of *Dracunculus medinensis*.

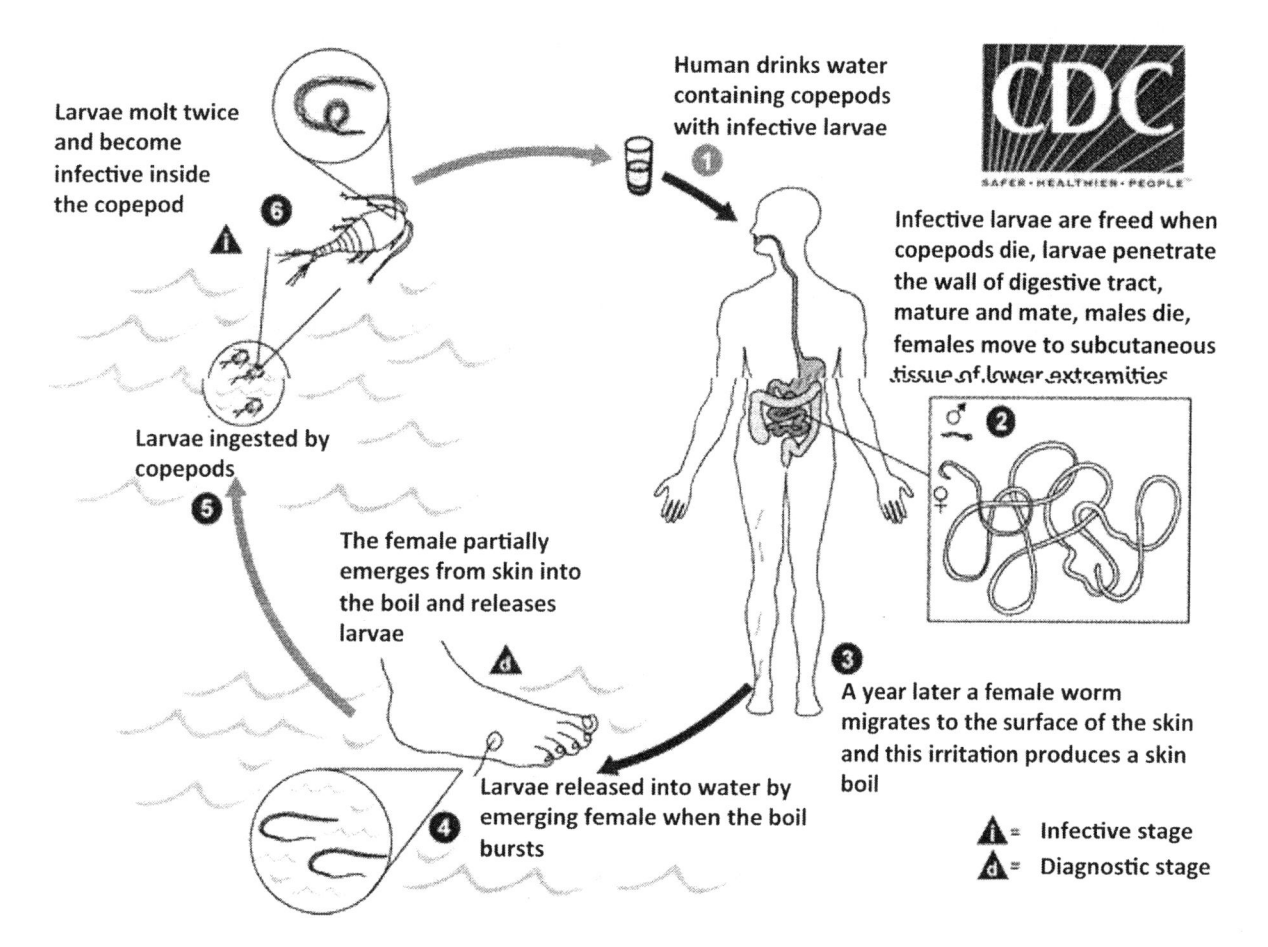

Figure 9.13. Life cycle of *Dracunculus medinensis,* the Guinea Fire Worm. A human is infected when they drink water that is contaminated with infected copepods (1), larvae burrows through the human's stomach or intestine wall and matures and mates (2), roughly a year later females now up to one meter long burrow into the subdermal tissues and then into epidermis, this irritates the skin and produces a fluid-filled blister or boil into which she extends the anterior part of her body (3), when the boil ruptures in water the female releases her larvae (4), larvae are ingested by copepods and larvae develop into infective stage larvae inside their copepod hosts (5-6). Filtering or boiling drinking water effectively breaks this cycle. With any luck these worms will soon be extinct globally. (Image: ARH modified CDC image, http://www.cdc.gov/parasites/images/guineaworm/dracunculiasis_lifecycle.gif)

5) *Dirofilaria immitis* – Dog heartworm. Lest this lab exercise come across as too anthropocentric, it's only fair to include a parasite that affects another species, in this case wolves and domesticated dogs though humans are also hosts to these worms. There are three species of *Dirofilaria* that all cause this disease but only *D. immitis* appears to affect domesticated dogs in North America. All of these are filarial roundworms that like *Wuchereria* move through the blood as microfilariae but in dogs the target tissue of adults are pulmonary arteries and sometimes the right ventricle of the heart, thus the common name of these worms. Fortunately these worms do not develop into adults in humans. **Figure 9.14** shows the life cycle of this parasite.

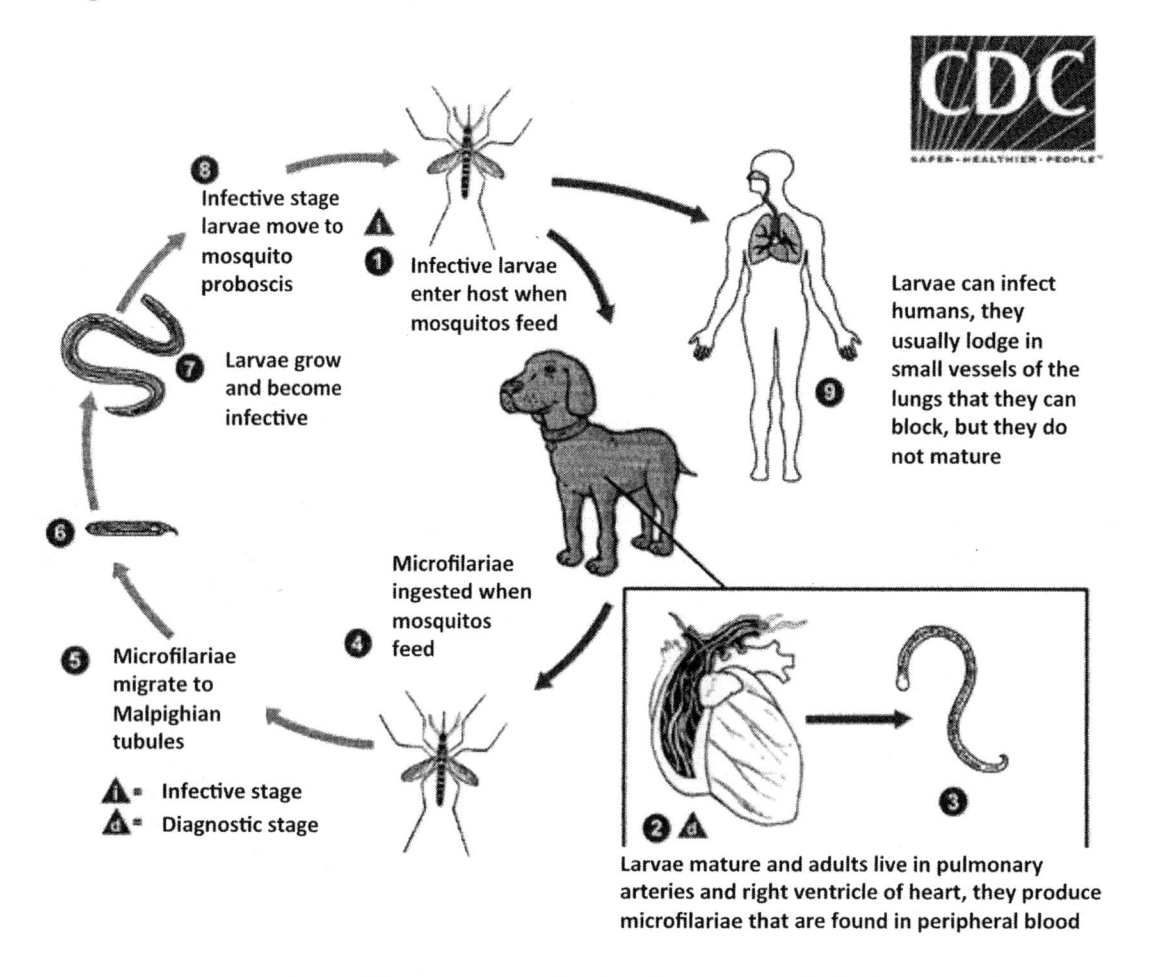

Figure 9.14. Life cycle of *Dirofilaria immitis*, the dog heartworm. Infective stage larvae are introduced into a host by a mosquito during feeding (1) and larvae move to the pulmonary arteries and sometimes right ventricle in dogs where the worms take up residence (2), adults release microfilariae (3) that are taken up by mosquitoes (4), and microfilariae develop into infective stage larvae in a mosquito's body, eventually making their way to the proboscis (5-8). Adult females reach lengths of 2-3 cm but males are smaller. According to the CDC adults can live up to 10 years with females producing microfilariae the entire time. Symptoms in dogs can include coughing up blood, blockage of vessels, fatigue and severe weight loss. Though these worms do not mature in humans, larvae migrate to blood vessels of the lungs where they can cause blockages and lesions. (Image: ARH modified CDC image, http://www.cdc.gov/parasites/dirofilariasis/biology_d_immitis.html)

Group Questions

1) Brachiopods were once extremely abundant, both numerically and taxonomically. Today however there are only about 350 living species of brachiopods, compared to 12,000 known fossil species. It has been suggested that the evolution of bivalve molluscs led to the decline in the dominance of brachiopods. Develop a hypothesis that explains how the appearance of bivalve molluscs caused a decline in brachiopod biodiversity and geographic distribution.
2) Go to the Centers for Disease Control website (www.CDC.gov) and learn about a roundworm parasite that is not included in this lab. Develop a plan for eradicating it.

Phylum Brachiopoda – Glossary

Abductor muscles – muscles that open the shell

Adductor muscles – muscles that close the shell

Anterior adjustor muscles – muscles attached to the anterior portion of the dorsal shell

Aperture of the ventral valve – opening where the pedicle passes through the shell

Articulate brachiopod – brachiopod with a tooth-and-socket hinge for attaching shells together

Beak of ventral valve – rounded to pointed end of the valve posterior to the aperture

Benthic – referring to the bottom in an aquatic environment

Brachial fold – lophophore tentacles are attached along this groove

Chaetae – bristle-like structures that keep large particles from entering the mantle cavity

Digestive cecum – organ used for digesting and absorbing nutrients

Hinge teeth – tooth-and-socket hinge used to hold shells together in articulate brachiopods

Inarticulate brachiopod – brachiopod that lacks a tooth-and-socket hinge on shells

Index fossil – a fossil that can be used to help date fossil-bearing rock layers

Lophophore – a ciliated crown of tentacles used for suspension feeding and gas exchange

Oblique adjustor muscles – muscles used by inarticulate brachiopods to keep shells aligned

Pedicle – an organ used to attach the animal to the substrate

Phylum Nematoda – Glossary

Blastocoelom – fluid-filled space derived from the embryonic blastocoel, a fluid-filled space in the body with an outer lining of mesodermal tissue and inner lining of endodermal tissue, also known as a pseudocoelom

Cloacal opening/Cloaca – an opening through which more than one organ system releases their products, e.g., feces and gametes

Common vagina – space just inside the female gonopore that leads to the uterus

Copepod – microscopic arthropod zooplankton

Cuticle – protective outer covering secreted by the epidermis, made of β-chitin in nematodes

Dorsal cord – thickened area of the body wall that houses a dorsal nerve cord

Ecdysis – process of molting the cuticle, regulated by the hormones ecdysone and molt-inhibiting hormone

Ejaculatory duct – tube that ejects sperm through the cloaca during copulation

Embryonated egg – embryo encased in an environmentally resistant outer covering

Female gonopore – opening through which sperm are received and embryonated eggs are released

Genital sensillum – small chemosensory extensions of the anal region of male *Ascaris*

Gravid – full of eggs or embryonated eggs

Larva – post-hatching life stage that has a different appearance and ecology than adults and is not sexually mature

Lateral epidermal cord – thickened section of the body wall that houses the lateral excretory canals

Microfilaria – microscopic stage of nematodes that is taken up by an insect while feeding

Mucosa – a mucous membrane, e.g., the lining of the intestine

Nerve process – extension of a longitudinal muscle that extends to either the dorsal nerve cord or the ventral nerve cord

Ovary – primary female reproductive organ, produces eggs

Oviduct – tube that carries eggs from the ovary to the uterus

Papillae – accessory structures used to hold the female during copulation

Pharynx – muscular structure between the mouth and intestine that pulls food into the body, is tri-radiate (Y-shaped) in nematodes

Pseudocoelom (see Blastocoelom)

Seminal vesicle – reservoir in the male body where sperm are stored in preparation for copulation

Sperm duct/vas deferens – tube between the testes and seminal vesicle

Sylvatic – term used to refer to diseases contracted by wild animals

Testes – primary male reproductive organ, produces sperm

Uterus – female organ where fertilized and unfertilized eggs are stored until they are released

Ventral cord – thickened area of the body wall that houses a ventral nerve cord

Chapter 10: Clade Panarthropoda

Clade Panarthropoda (means "all jointed legs") includes the following three phyla: Tardigrada, Onychophora and Arthropoda. These phyla have segmented bodies or evidence of segmentation, paired appendages and a non-elastic outer cuticle or exoskeleton that is molted periodically via ecdysis. Tardigrada is currently accepted as the basal taxon in this clade and as the sister taxon to Onychophora-Arthropoda group.

Phylum Tardigrada "slow stepping" – water bears or moss piglets

There are about 1200 described species of tardigrades. Tardigrades consistently go unseen and unappreciated because they are small, obscure and apparently medically and ecologically unimportant. Most of these animals are only 0.1-0.5 mm long but a few species can exceed 1.0 mm in length. Tardigrades are so tiny that some live in thin films of water on the surfaces of moist mosses and other plants as well as in aquatic environments. Tardigrades make a living by sucking the cytoplasm out of plant cells after piercing them with a sharp stylet that they secrete. Characteristics of tardigrades are listed in **Table 10.1**.

Table 10.1. Characteristics of Phylum Tardigrada (after Brusca, *et al.*, 2016).

- Evidence of segmentation
- 4 pairs of non-jointed telescoping legs that bear up to a dozen claws or pads per leg
- Non-calcified cuticle that is molted periodically via ecdysis
- Muscles in isolated bands
- Body coelom is a hemocoel
- Malpighian tubules
- Stylet for piercing cells
- Anabiosis/cryptobiosis

You can often find tardigrades yourself by collecting wet mosses growing in seldom-cleaned rain gutters, bird baths, finely branched aquatic plants or samples from moist cracks in tree bark where lichens are growing. These animals are also available from biological supply companies. Watch this short video to see what to look for in your samples: https://www.youtube.com/watch?v=aHsVyb_VfeA

Tasks – Phylum Tardigrada

1) Observe live tardigrades. If you collect materials where tardigrades may exist, soak the material in 5-10% ethyl alcohol solution. After 5-15 minutes in the solution gently shake the moss or aquatic plant. If tardigrades are present they should be relaxed by this point and drift to the bottom of the container. Use a pipette to collect samples from the bottom of the jar or bowl and make a wet mount slide. Put sufficient plasticene clay on the corners of your coverslip so that your specimens are not smashed. Tardigrades will be the only thing in your sample with 4 pairs of lobe-shaped appendages that bear chitinous hooks as shown in **Fig. 10.1**. When you find one get excited, show your instructor, show

your neighbor and then DRAW what you see. Be sure to WRITE plenty of observations about their behavior, refer to **Fig. 10.1** to identify tardigrade anatomy.

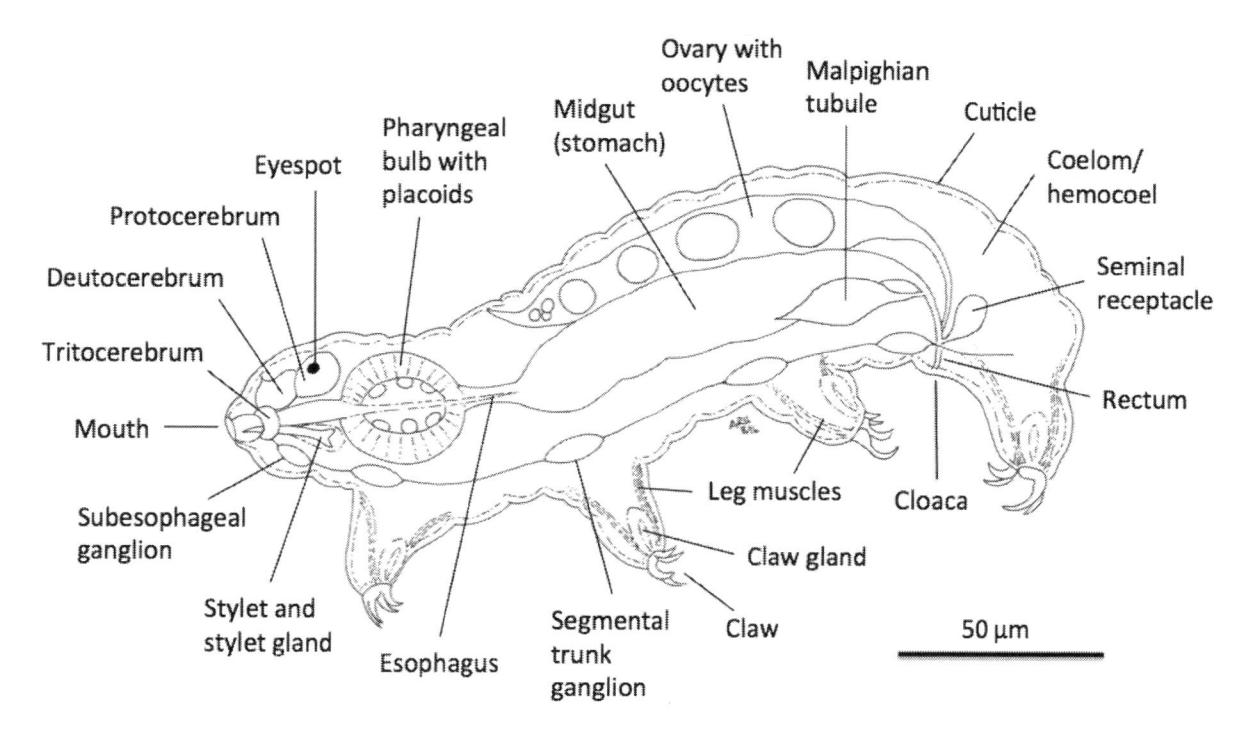

Figure 10.1. Tardigrade anatomy, only one set of appendages is shown. (Image: ARH)

2) Tardigrade eggs or embryos. Female tardigrades routinely release eggs into their exoskeletons as they undergo ecdysis. These eggs may already be fertilized or fertilization can occur afterwards. Scan your sample to see of you can find any eggs or developing embryos in otherwise empty molts. DRAW what you see. Use **Fig. 10.2** to see an example of developing tardigrade embryos in a female molt.

Figure 10.2. Tardigrade embryos in parental molt. Note the parental cuticle and hooks (Image: Courtesy of Dr. Frank E. Anderson, Dept. of Zoology, Southern Illinois University – Carbondale)

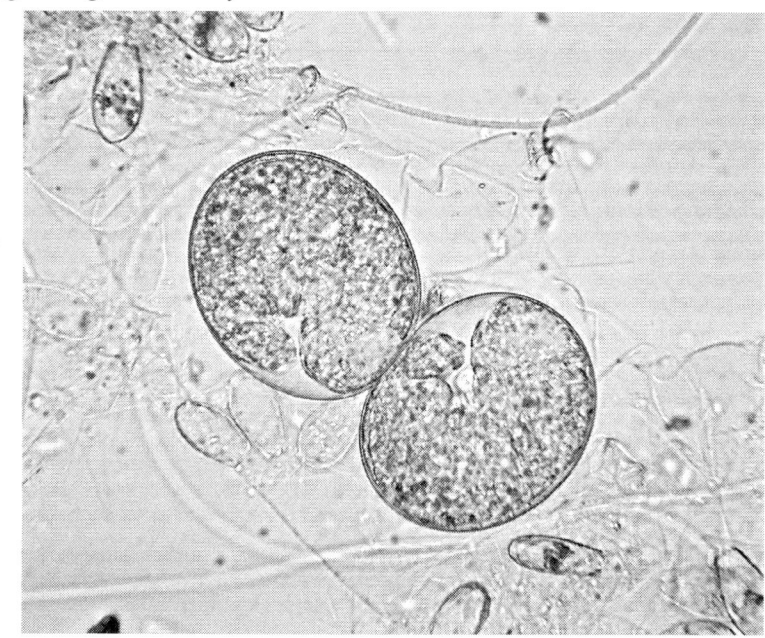

3) Tardigrade anabiosis. Tardigrades are able to survive extreme environmental conditions including desiccation, temperature extremes and even the complete vacuum of space and prolonged exposure to full spectrum sunlight. They can do this when they are in their tun stage, a condition of anabiosis or suspended animation that they enter when they become completely desiccated. As they dry out tardigrades withdraw their legs and shrivel up. When you are done observing live tardigrades collect several tardigrades and place them on a microscope slide or in a small bowl or watch glass with as little water as possible. Set the specimens aside to dry. The next time you are in lab look at your specimens and find the desiccated animals – these may now be in the tun stage. DRAW what you see, they will most likely look like little shriveled dots that look like mini-raisins. Add water to the specimens at the beginning of your next laboratory period and observe them every 5-10 minutes to see if they will rehydrate and become reanimated. WRITE what you see in your laboratory notebook about desiccation and reanimation.

Phylum Onychophora "claw bearing" – velvet worms or walking worms

Onychophorans have an interesting distinction; this is the only animal phylum that is strictly terrestrial, though their ancestors were marine species that lived as long as 500 million years ago. The current geographic range of living species includes only tropical or subtropical regions of the Neotropics (the Americas), Africa and Asia. These animals require high humidity to survive because they cannot close the spiracles of their tracheal systems, this can result in rapid desiccation when they experience dry conditions. Characteristics of this group are listed in **Table 10.2**.

Table 10.2. Characteristics of Phylum Onychophora (after Brusca, *et al.*, 2016).

- Segmented body
- Telescoping, non-jointed legs (lopopods) that bear chitinous hooks
- Muscles in distinct bands, plus a sheet of muscles in the body wall
- Jaws
- Oral/slime papillae
- Trachea and spiracles
- A paired ventral nerve cord lacking any evidence of segmentation

Onychophorans are nocturnal and though they are slow movers that are practically blind they are effective hunters. Their unique hunting strategy allows them to capture even relatively large prey. Velvet worms locate their prey by sensing micro-currents of air produced by potential prey. Velvet worms shoot long streams of sticky slime that immobilizes their prey just like Spiderman shoots web to immobilize bad guys except Spiderman doesn't eat them; onychophorans on the other hand are voracious predators. These video clips show velvet worms in action. https://www.youtube.com/watch?v=3DOvo2V8XIY
https://www.youtube.com/watch?v=mrL2A7my1fc

Phylum Onychophora - Taxonomy

Onychophorans were once viewed as an evolutionary link between annelids and arthropods. Modern taxonomic analysis reveals however that onychophorans are not a missing link but instead are the sister taxon of arthropods; they are only very distantly related to annelids.

There are two families of onychophorans. The main distinction between these groups has to do differences in jaw structure, the number of pairs of appendages they produce and their geographic ranges but additional details of onychophoran taxonomy are not addressed here.

Tasks – Phylum Onychophora

1) External anatomy of *Peripatus*. Use a magnifying glass or dissection scope to examine a preserved onychophoran. DRAW the basic body plan and use **Fig. 10.3** to help you identify what you see.

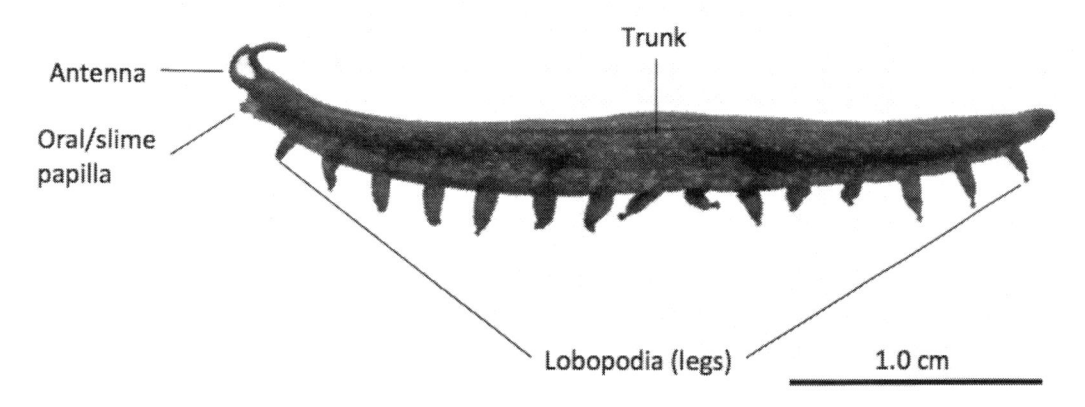

Figure 10.3. The onychophoran *Peripatus*. (Image: ARH)

2) Use a magnifying glass or dissection scope to examine the head, lobopods (appendages) and posterior end of the body. Look for eyespots at the base of the tentacles and nephridiopores at the base of the lobopods. These are difficult to see in preserved specimens but check anyway. In addition look for the anus posterior to the last set of lobopods and the genital opening that will be somewhere along the ventral midline between the last set of lobopods or slightly posterior to them. DRAW what you see. Refer to **Fig. 10.4** to help you identify structures of your specimen.

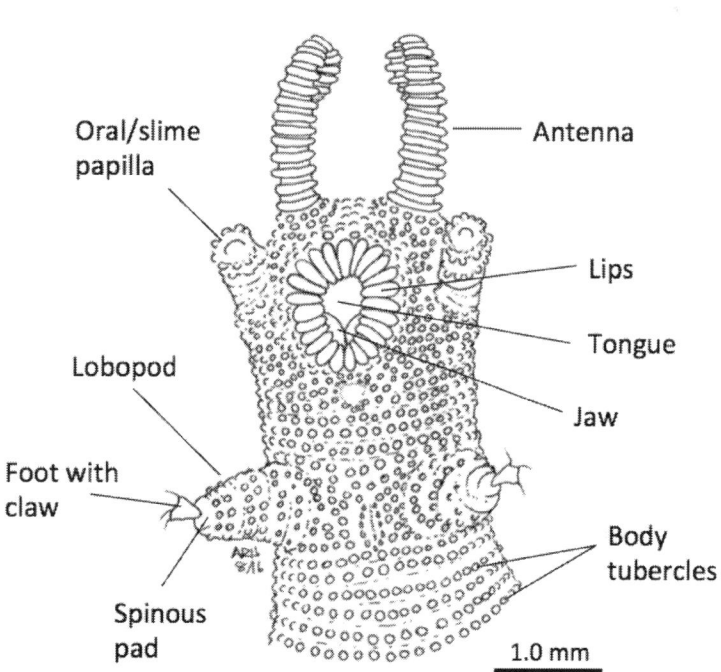

Figure 10.4. Head and first pair of lobopods of *Peripatus*, **ventral view**. (Image: ARH)

Phylum Arthropoda "jointed foot" – insects, crustaceans, arachnids and relations

There are over one million described species of Arthropods making it the largest animal phylum in terms of described species. Most arthropods are insects. Phylum Arthropoda is in terms of number of species and ecological significance the most successful phylum of animals. Arthropods occupy virtually every life-supporting habitat on the planet including marine, freshwater and terrestrial environments, there are even arthropods in Antarctica. Their basic body plan has conveyed this enormous success and their characteristics are listed in **Table 10.3**.

Arthropods have a segmented body and a protective outer cuticle or exoskeleton. Segments of the body are typically fused together to produce specialized body regions called tagma, e.g., head, thorax, abdomen, and their jointed appendages are specialized to carry out a wide diversity of specific tasks as depicted in these video clips from the series *The Shape of Life*: http://shapeoflife.org/video/marine-arthropods-successful-design, http://shapeoflife.org/video/terrestrial-arthropods-conquerors.

Table 10.3. Characteristics of Phylum Arthropoda (after Brusca, *et al.*, 2016).

- Segmented body typically including tagma
- Cuticle or exoskeleton periodically replaced via ecdysis
- Jointed appendages
- One pair of appendages per body segment (ancestral condition)
- Paired ventral nerve cord with segmental ganglia
- Segmental muscles
- Simple ocelli or compound eyes or both (in most)

Phylum Arthropoda – Taxonomy

Phylum Arthropoda includes five subphyla, four that are extant and one that is extinct.

Subphylum Trilobitomorpha/Trilobita "three lobes" – trilobites, extinct. Trilobites were prominent members of marine communities globally 500-225 million years ago and produced many index fossils. The last trilobites went extinct during the Permian Extinction also called "The Great Dying," when 96% of all marine species went extinct.

Subphylum Chelicerata "clawed horn" – horseshoe crabs, sea spiders, ticks, scorpions, spiders and relations. Most chelicerates have two tagma, a prosoma and an opisthosoma. The prosoma bears a pair of uniramous chelicerae, pedipalps and four pairs of walking appendages but lacks antennae. The brain has three lobes and there are simple and compound eyes. Gas exchange organs vary between groups but include book gills, book lungs and trachea. Chelicerates use Malpighian tubules for excretion.

Subphylum Myriapoda "numberless feet" – centipedes, millipedes and relations. This body plan has a head and a trunk. The head bears one pair of antennae, mandibles and two pairs of maxillae. The trunk is made of many identical segments that bear uniramous appendages. The brain has two lobes and most species have only simple eyes. Myriapods also have Malpighian tubules.

Subphylum Crustacea "crust or hard outer covering" – shrimp, lobsters, crabs, barnacles and relations. This group contains ecologically and economically important species. They are found in marine, freshwater and terrestrial environments. There are crustaceans that are predators, herbivores, scavengers, suspension feeders and parasites. These animals usually have a cephalothorax and abdomen and two pairs of antennae.

Subphylum Hexapoda "six feet" – insects and relations. Hexapods probably evolved within the Crustacea but they continue to be treated independently for the time being, so don't be surprised if hexapods lose their subphylum status sometime in the near future. All hexapods have a head, thorax and abdomen. Four pairs of appendages are associated with the head: one pair of antennae, mandibles, maxillae and labrum. The thorax has three segments, each of which bears one pair of uniramous legs. Hexapods also have a three-part brain, compound eyes, spiracles and trachea and Malpighian tubules. In addition insects typically have two pairs of wings. Insects are the only group of animals that evolved wings without sacrificing a pair of walking appendages to make them.

Subphylum Trilobitomorpha / Trilobita – trilobites

Though we focus primarily on living taxa in this lab manual every student of zoology should at least know about trilobites. Characteristics of trilobites are listed in **Table 10.4**.

Table 10.4. Characteristics of Subphylum Trilobitomorpha (after Pechenik, 2015).

- Bodies divided into three tagma: cephalon, thorax and pygidium
- The body has three longitudinal lobes: a central axial lobe and left and right pleural lobes (these lobes give trilobites their name)
- Segmental biramous appendages
- Compound eyes
- One pair of antennae

This short animated video, though not particularly scientific, does a pretty good job of depicting some of the things we think living trilobites were able to do.
https://www.youtube.com/watch?v=znO8q5Ht17g

Tasks – Subphylum Trilobita/Trilobitomorpha

1) Examine a trilobite fossil and DRAW what you see. Use **Fig. 10.5** to help you identify trilobite anatomy.
2) Visit a Geology or Natural History museum on your campus or in your community (if available) to peruse the diversity of trilobites as well as other fossils on display. WRITE observations of your visit to the museum in your lab notebook. This visit will probably need to take place outside of scheduled class or lab time.

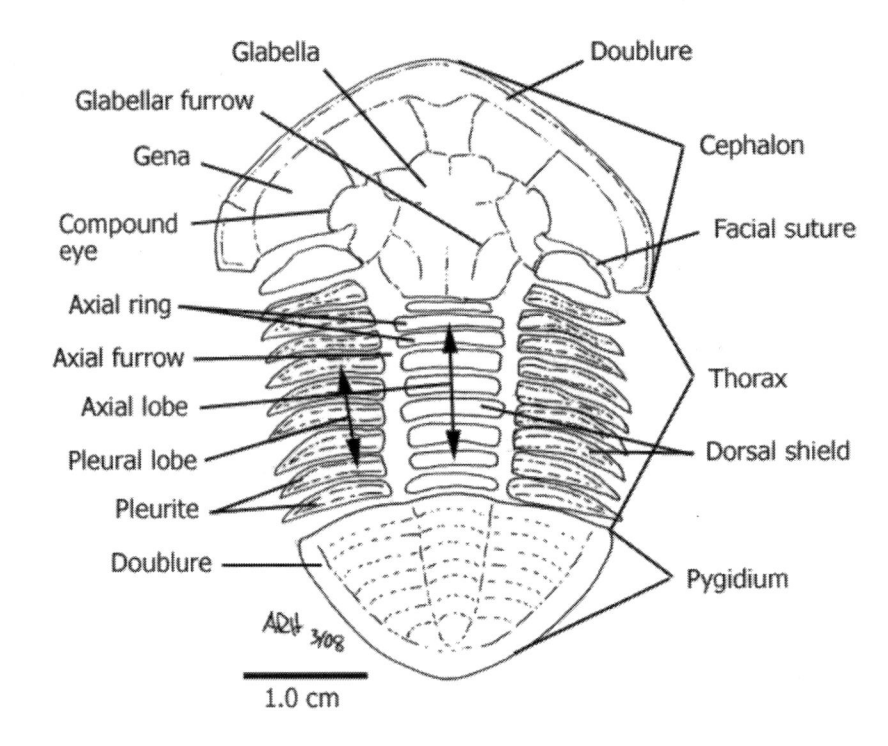

Figure 10.5. Trilobite, dorsal view. (Image: ARH)

Subphylum Chelicerata

Most chelicerates are terrestrial and include scorpions, spiders, mites and their relations. Marine forms include horseshoe crabs and sea spiders. Characteristics of chelicerates are listed in **Table 10.5**.

Table 10.5. Characteristics of Subphylum Chelicerata (after Brusca, *et al.*, 2016).

- Body with two tagma: prosoma (6 segments) and opisthosoma (up to 12 segments)
- Telson attached to the opisthosoma present in some groups, is not a true body segment
- Appendages, anterior to posterior
    - Chelicerae
    - Pedipalps
    - 4 pairs of walking legs
- No antennae
- Gas exchange via books lungs, book gills, trachea or cuticle
- Malpighian tubules and other structures for excretion

Treatment of this group focuses on the external anatomy of chelicerate diversity: sea spiders, horseshoe crabs, ticks, scorpions and spiders.

Class Pycnogonida "Thick knees" – Sea spiders

Taxonomic work on pycnogonids shows that they are clearly arthropods but are only distantly related to terrestrial spiders. Pycnogonids are most likely the sister taxon to the rest of the chelicerates.

Most sea spiders are so small that the leg span of most species is small enough that they could fit easily on one of your fingernails. Some deep-sea species are relative giants with leg spans of up to 60 cm. Pycnogonids make a living by eating algae or a wide variety of invertebrates especially sea anemones and hydroids or tunicates. When pycnogonids eat they slice a hole in the body wall of their prey, insert the proboscis and use suction to pull food into their body. This video shows a deep sea pycnogonid:
https://www.youtube.com/watch?v=EK4KxKNqBbQ

Tasks – Class Pycnogonida

1) Observe dorsal and ventral views of a preserved pycnogonid and DRAW what you see. Also, DRAW one leg in detail and label the names of the articles that make up the leg. Refer to **Fig. 10.6** to help you identify these structures.

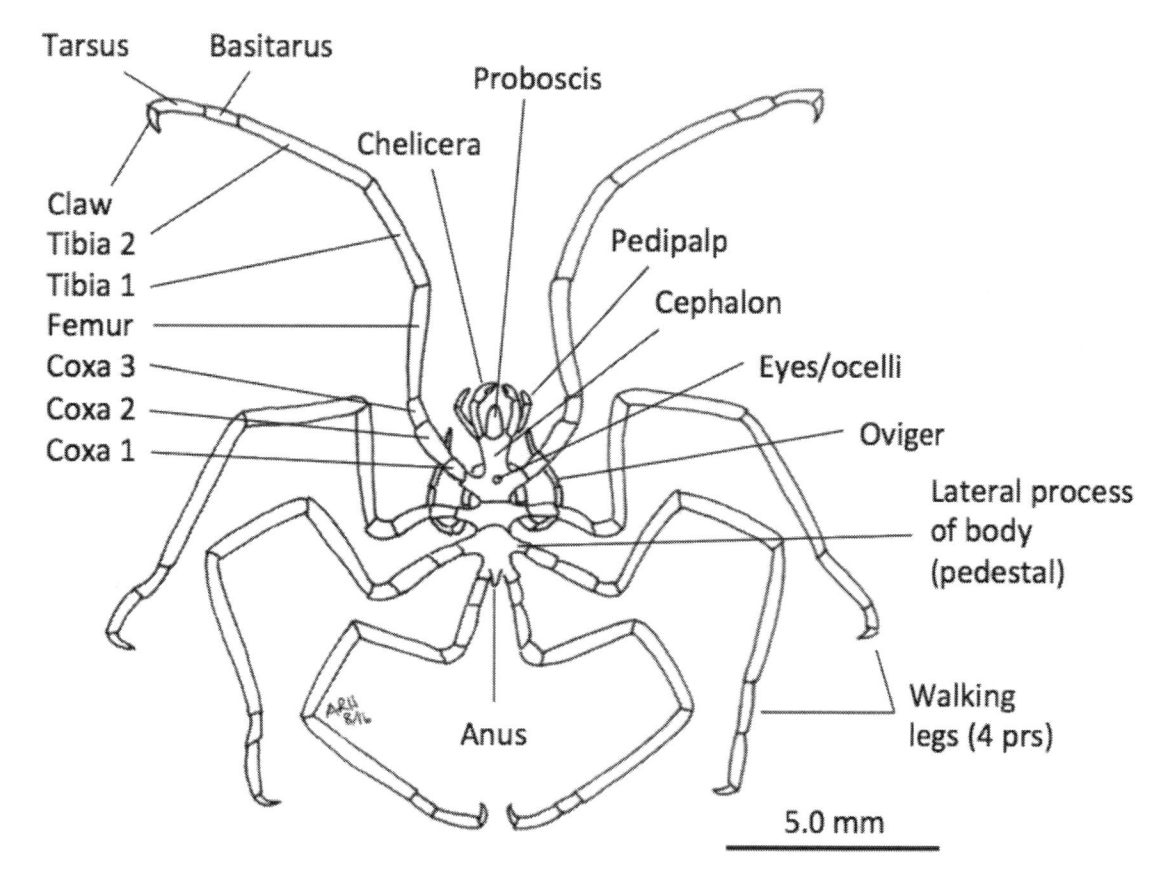

Figure 10.6. Anatomy of the pycnogonid *Nymphon brevirostre*, **dorsal view**. (Image: ARH, after D. W. Thompson in Pycnogonida, The Cambridge Natural History, Vol. IV, 1920)

Class Euchelicerata "true clawed horn", Order Xiphosura "sword above" – Horseshoe crabs

Horseshoe crabs are living fossils. Fossils similar to living species date back about 450 million years. There are only a four living species of horseshoe crabs. One of these is *Limulus polyphemus* a species that is economically important because it is the source of limulus amoebocyte lysate, a substance that can sell for around $15,000/liter and is used in biochemical and biomedical research. Watch these short videos about horseshoe crabs to learn more about them and limulus amoebocyte lysate.
https://www.youtube.com/watch?v=90LTtKIFY8U
https://www.youtube.com/watch?v=-55qGCHx11E

Horseshoe crabs spend most of their lives feeding in deeper water but when it is time to reproduce they crawl onto beaches during a spring high tide. Mating occurs there and females dig nests at the high tide line where they deposit fertilized eggs. Eggs incubate in the warm sand and hatch about a month later during the next spring high tide. North American horseshoe crab populations are threatened by shoreline development and in some areas by eel and conch fisheries that use these animals for bait.

Tasks – Horseshoe crabs

1) Horseshoe crab larva. Use a dissection or compound microscope to examine a prepared slide of the trilobite larva of *Limulus* and compare the larva to an adult. DRAW the larval stage and use **Fig. 10.7** to help you identify what you see.
2) External anatomy. Study external anatomy of dorsal and ventral surfaces of adult *Limulus*. DRAW what you see. Refer to **Fig. 10.8** to help you identify what you see.

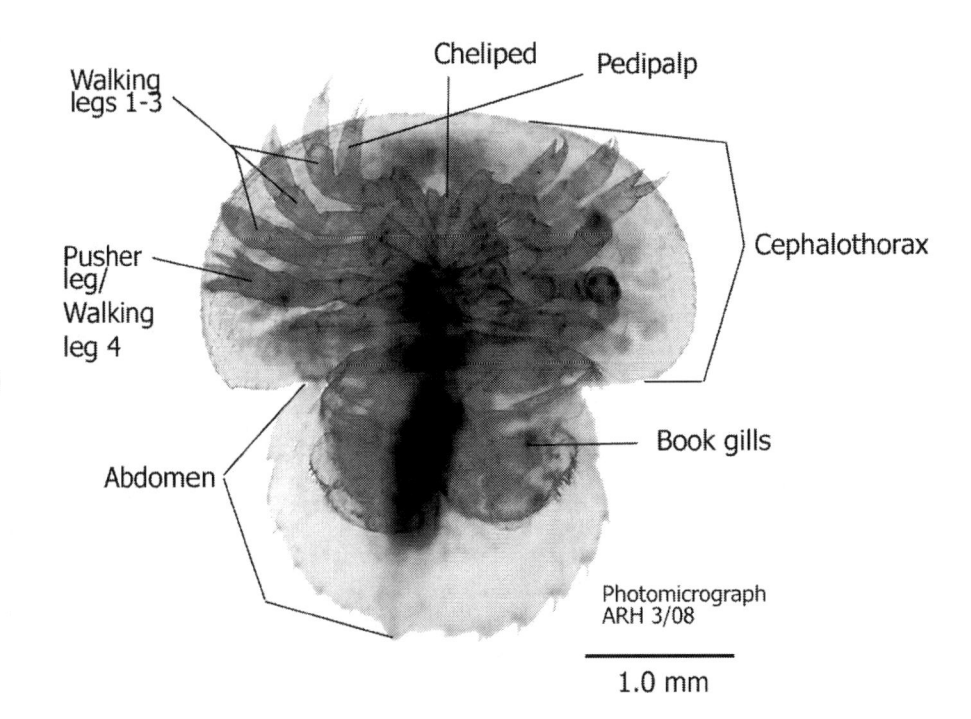

Figure 10.7. Trilobite larva of *Limulus*, ventral view. (Image: ARH)

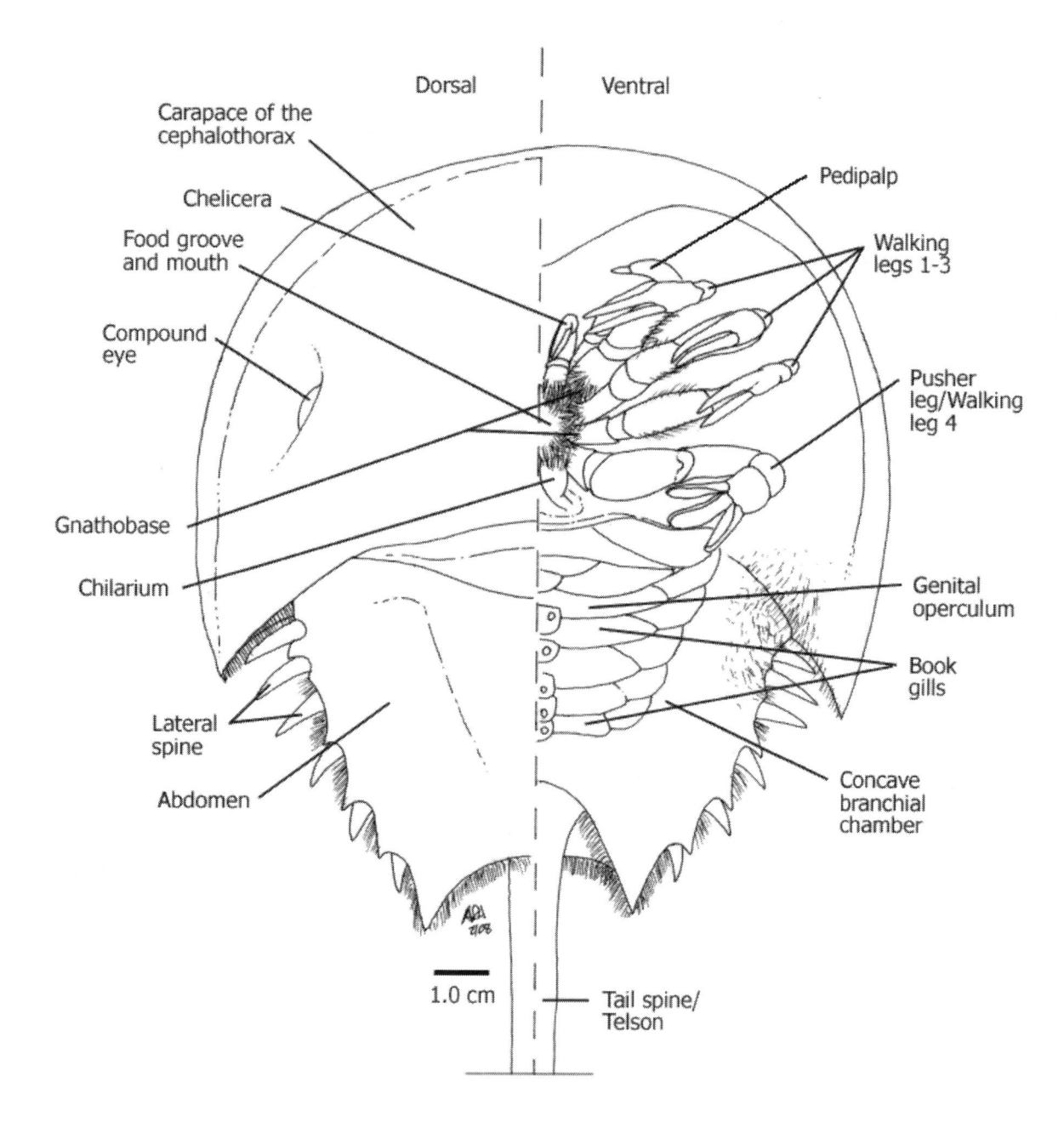

Figure 10.8. External anatomy of *Limulus,* dorsal surface, **left half**; and ventral surface, **right half**. (Image: ARH)

Subclass Arachnida "a spider" – ticks, scorpions, spiders and relations

Subclass Arachnida is a large taxon of more than 110,000 mainly terrestrial species. Living members of this taxon include ticks, mites, scorpions, spiders, etc. Terrestrial arachnids use book lungs, trachea, the cuticle or a combination of these to carry out gas exchange. They also use Malpighian tubules for excretion and produce uric acid, a dry form of nitrogenous waste that helps these animals conserve water.

In this portion of the exercise you are introduced to the diversity of arachnids by examining the external anatomy selected representatives.

These short videos do a good job of showing aspects of the biology of ticks, scorpions and spiders (in that order):
https://www.youtube.com/watch?v=0g_lt0FcQag
https://www.youtube.com/watch?v=DUz7YRhqick
https://www.youtube.com/watch?v=ysUZ6mbFQ3w

Tasks – Tick
1) Use a dissection microscope or compound microscope to study a prepared slide or other available specimen of a tick. DRAW the body plan of your specimen and use **Fig. 10.9** to help you identify what you see.
2) Use a compound microscope to study the structure of the capitulum. DRAW what you see and use **Fig. 10.10** to help you identify the anatomy of this part of the body.

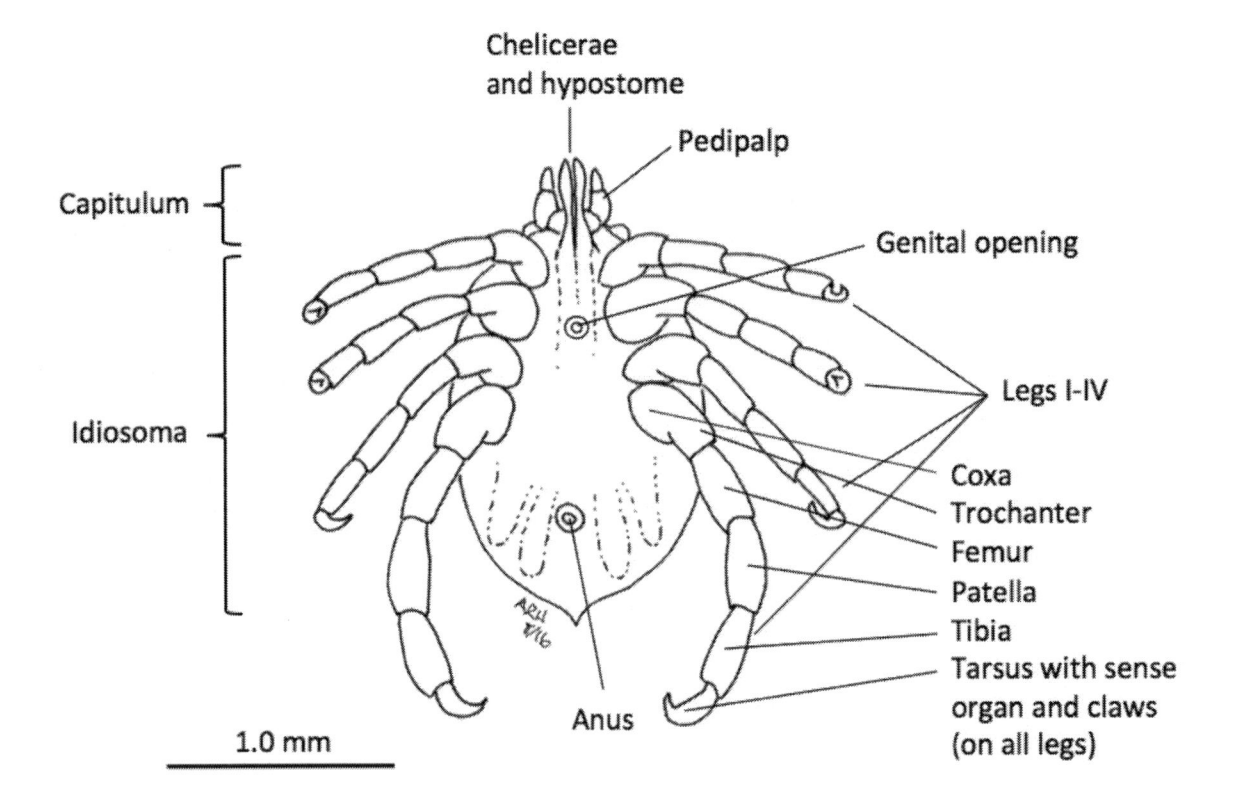

Figure 10.9. External anatomy of *Rhipicephalus* (*Boophilus*) *annulatus*, the blue cattle tick, **ventral view**. (Image: ARH)

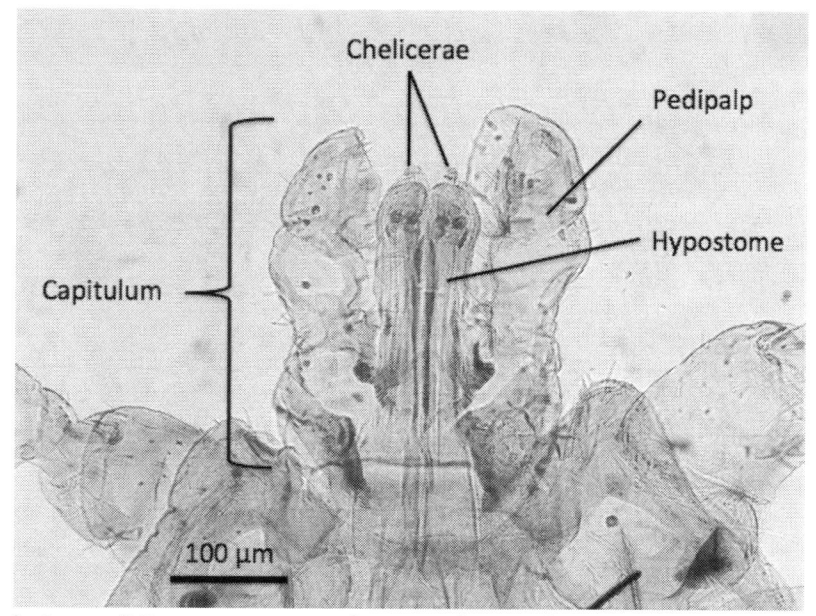

Figure 10.10. Capitulum of *Dermacentor andersoni*, the Rocky Mountain wood tick (Image: ARH)

Tasks – Scorpion

1. Use a magnifying lens and dissection scope to examine the external anatomy of a scorpion. DRAW the scorpion body plan and use **Figs. 10.11-12** to help you identify the anatomy of your scorpion.

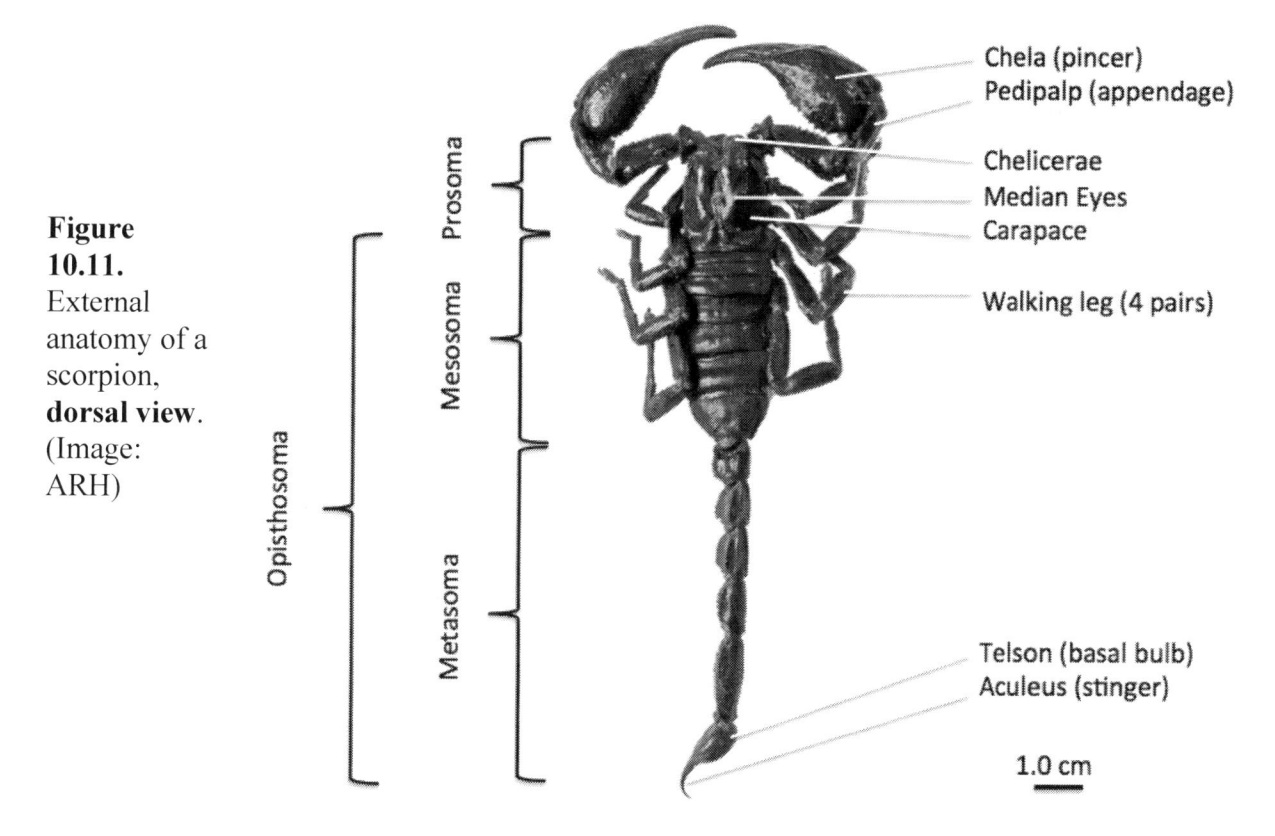

Figure 10.11. External anatomy of a scorpion, **dorsal view**. (Image: ARH)

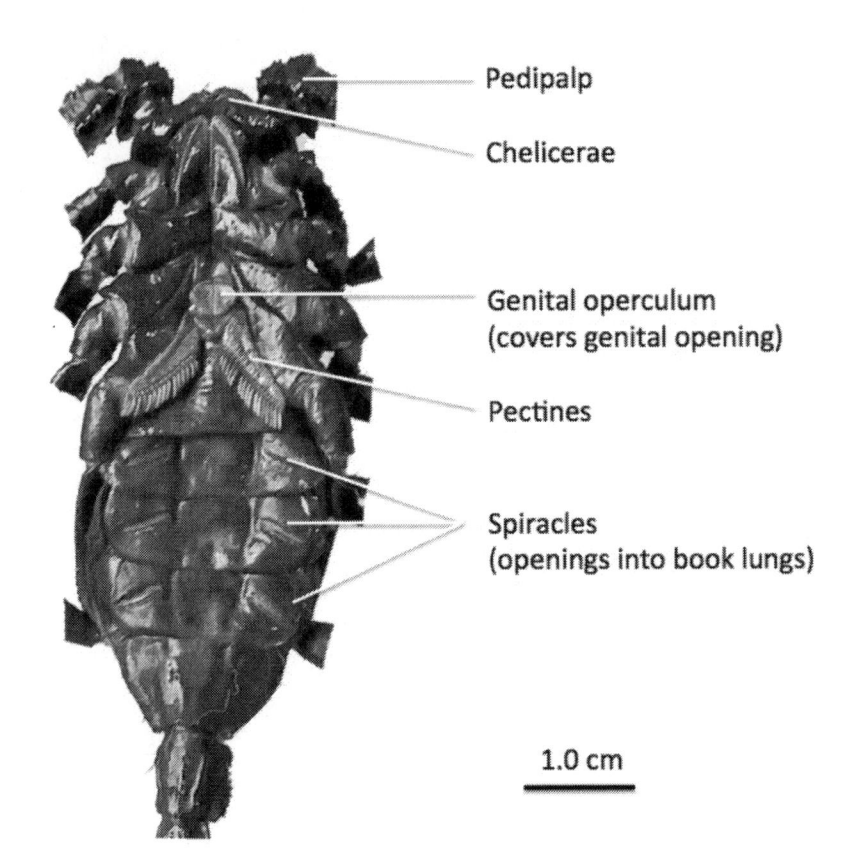

Figure 10.12. Prosoma and mesosoma of a scorpion, **ventral view**, appendages and metasoma not shown for simplification. (Image: ARH)

Tasks – Spider

1. Examine the external anatomy of a preserved spider. Use a magnifying glass or dissection scope as needed. Be particularly careful while handling your spider because they are notoriously fragile and appendages break off easily. DRAW what you see and use **Fig. 10.13** to help you identify the anatomy of your spider.
2. Observe the movement and behavior of a living spider if available. Pay particular attention to how they move opposing legs and support the body as they walk. WRITE your observations in your lab notebook.

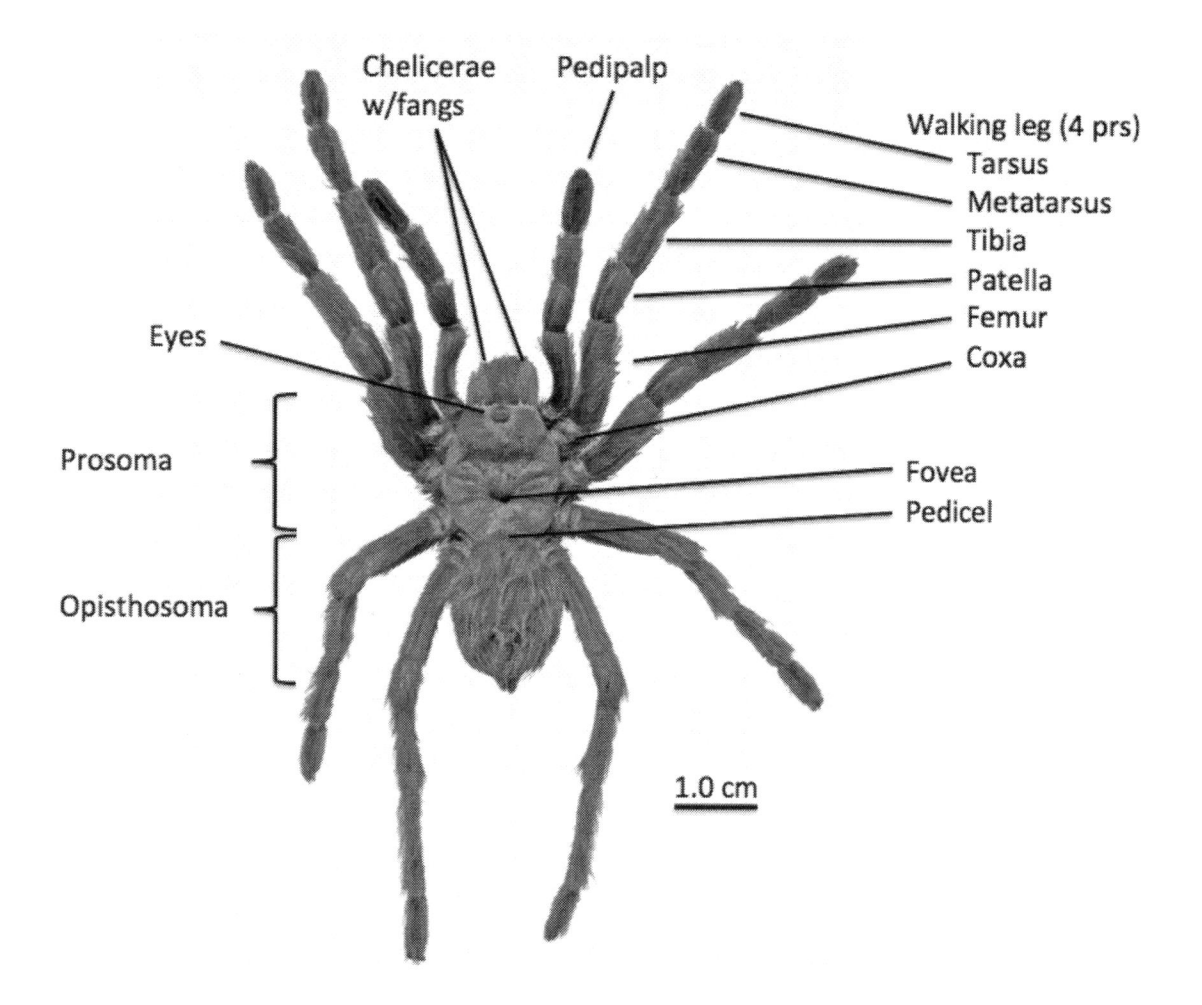

Figure 10.13. External anatomy of a tarantula, **dorsal view.** (Image: ARH)

Subphylum Myriapoda "numberless legs" – centipedes and millipedes

Myriapods include the centipedes, millipedes and a few other small groups not described here. All 13,000 species of myriapods are terrestrial. Characteristics of myriapods are listed in **Table 10.6**.

Table 10.6. Characteristics of Subphylum Myriapoda (after Brusca, *et al.*, 2016).

- Body with two tagma – head and multisegmented trunk
- Each body segment with dorsal tergite, ventral sternite and lateral pleurite exoskeletal plates
- All appendages have several articles and are uniramous
- Head appendages
    - One pair of antennae
    - Mandibles
    - Two pairs of maxillae
- Trachea and spiracles for gas exchange
- Malpighian tubules for excretion

Class Chilopoda "thousand feet" – centipedes

The name centipede means "100 feet" but ironically all centipedes have an odd number of leg-bearing body segments so no centipede has 100 feet. Centipedes have one pair of legs per trunk segment and can run quickly. They are fast because their legs are long, are attached to the sides of each body segment and one leg of a pair swings forward while the other leg pushes backward in long arcs giving them long steps. They undulate from side-to-side as they move.

Centipedes are active hunters. They subdue their prey by grasping and injecting poison into them via maxillipeds modified into structures variously referred to as prehensors, forcipules or fangs.

This video shows some aspects of the behavior of centipedes: https://www.youtube.com/watch?v=vutTYCBlA5M

Tasks – Class Chilopoda

1) Examine the external anatomy of a centipede. Look carefully at the head and legs of your specimen. Observe the anatomy of dorsal and ventral surfaces and the posterior end of the body. DRAW your specimen and use **Figs. 10.14-15** to help you identify what you see.

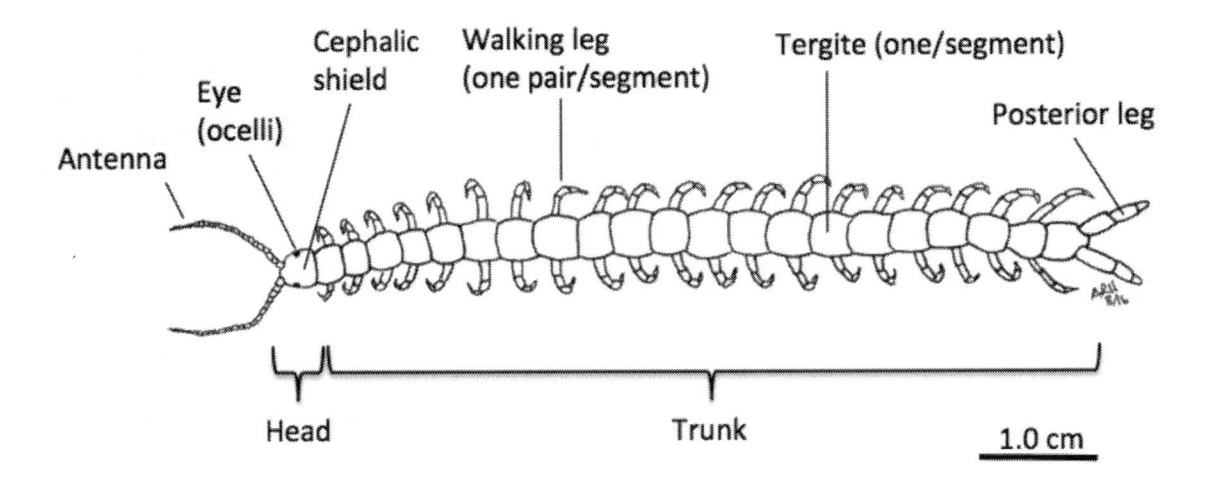

Figure 10.14. External anatomy of the centipede *Scolopendra*, **dorsal view**. (Image: ARH)

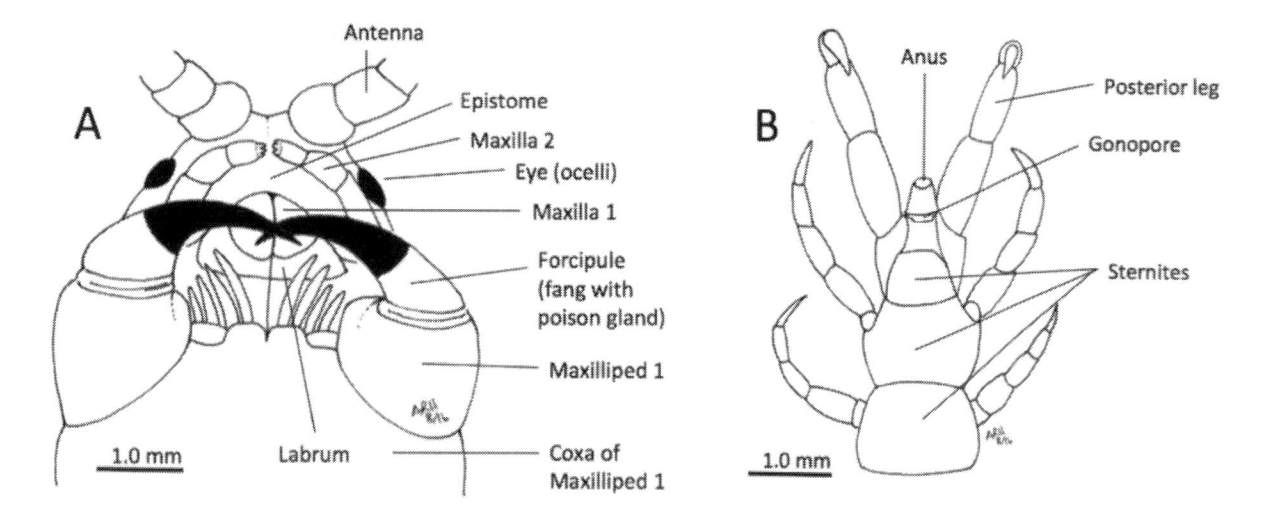

Figure 10.15. Anatomy of the **(A)** head and **(B)** posterior end of *Scolopendra*. (Images: ARH)

Class Diplopoda "double feet" – millipedes

Millipedes are slow-moving herbivores and detritovores. Like centipedes, the millipede body has a head and trunk but millipede bodies are nearly circular in cross-section while those of centipedes are dorso-ventrally compressed. In millipedes the first three to five segments bear only one pair of appendages each, all other segments of the trunk are diplosegments. A diplosegment is made of two fused segments and bears two pairs of appendages, two pairs of spiracles, etc. Millipedes are slow movers. Their legs are short, attached near the ventral midline and each pair of legs steps forward and backward together instead of in alternate step as in centipedes. Though slower, the millipede stepping motion together with their large numbers of legs allows them to push/burrow/wedge themselves into places where they need to go.

Millipedes have effective defensive strategies. The exoskeleton is calcified and millipedes can coil up. They also have a chemical defense; they produce an alkaloid compound that is in the same chemical class as methaqualone, marketed as Qualuudes – a CNS depressant and hypnotic sedative. Imagine if a spider attacks a millipede the millipede can releases its chemical defense and the spider almost immediately feels relaxed and so good that it forgets what it was doing and the millipede escapes.

This video is not scientific but it shows the defensive capabilities of millipedes: https://www.youtube.com/watch?v=UxuayPr0SGs

Task – Class Diplopoda

1) Examine the external anatomy of a millipede. Be careful while handling preserved millipedes because they are quite fragile and readily break in half. Look closely at structures of the head, the first 3-5 trunk segments and at diplosegments as well as the posterior portion of the body. DRAW your specimen and use **Figs. 10.16-17** to help you identify what you see.

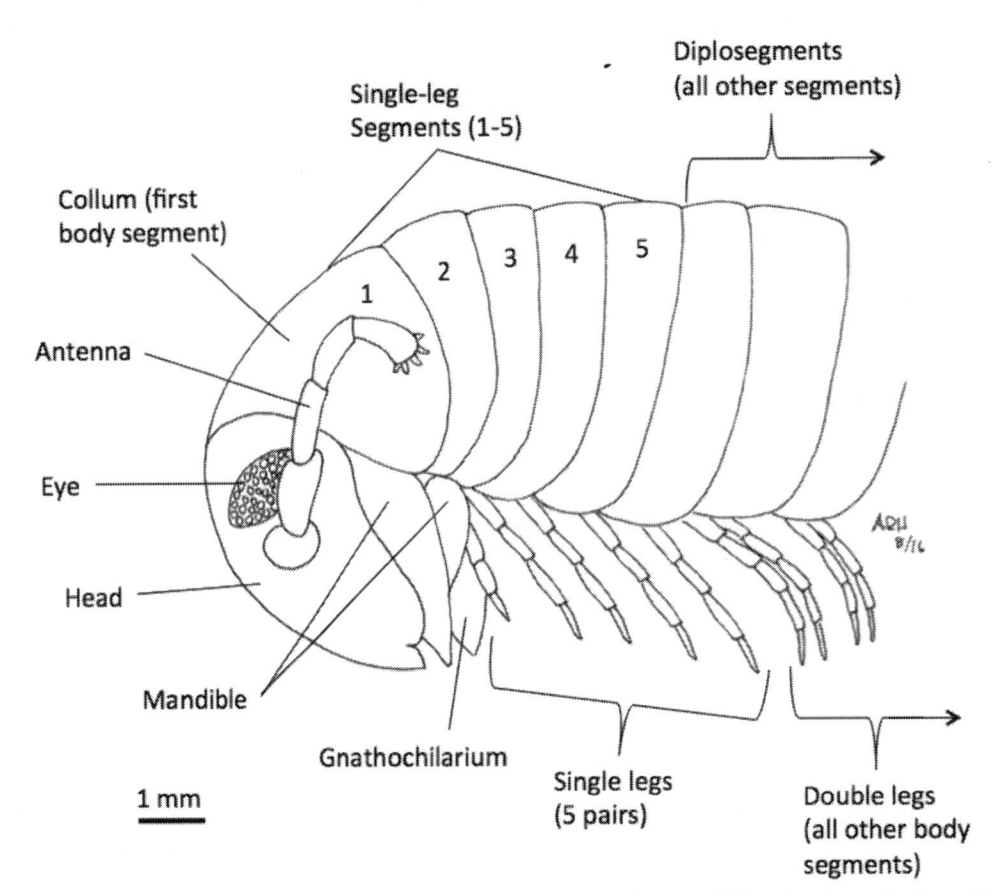

Figure 10.16. Anatomy of the head and anterior portion of the millipede *Spirobolus*. The first five trunk segments of *Spirobolus* bear a single pair of appendages and all other trunk segments are diplosegments. (Image: ARH)

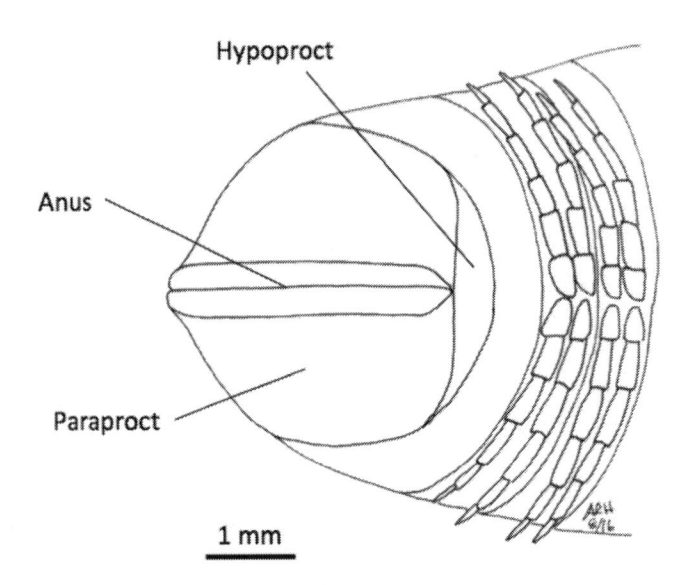

Figure 10.17. Anatomy of the posterior end of the millipede *Spirobolus*. There is also a plate called the hyperproct (not shown) located dorsal to the anus. (Image: ARH)

Subphylum Crustacea "crust or rind group" – water fleas, shrimp, crabs and relations

Subphylum Crustacea is an ecologically and economically important taxon. Crustaceans are familiar to most people because this group contains lobsters, crabs and shrimp. The vast majority of crustaceans live in marine and freshwater habitats though a few species have adapted to terrestrial life. One group of familiar terrestrial crustaceans are known by various common names, e.g., woodlouse, pill bug, potato bug, armadillo bug, sow bug, roly-poly, and a new one to me, chucky pig – a term reportedly used in Devon, England.

Crustaceans include predators, herbivores, omnivores, suspension-feeders, scavengers and even parasites. Characteristics of crustaceans are listed in **Table 10.7**.

Table 10.7. Characteristics of Subphylum Crustacea (after Brusca, *et al.*, 2016).

- Body with a 6-segment head and a trunk usually divided into a thorax and an abdomen
- Cephalic appendages:
    - Two pairs of antennae (antennae and antennules)
    - Mandibles
    - Two pairs of maxillae
- Multiarticulate uniramous or biramous appendages
- Gas exchange via gills or the cuticle
- Simple ocelli and compound eyes
- Nauplius larva (found only in this taxon, though not all crustaceans produce nauplii)

Tasks – *Daphnia*, the water flea

Daphnia is a prominent member of the zooplankton in freshwater lakes and ponds. These animals belong to a group called cladocerans and they provide an important ecological link between smaller zooplankton and planktivorous fishes such as sunfishes, kokanee (sockeye salmon), etc. Specimens can be obtained relatively easily from open water by carrying out plankton tows from a boat or dock.

1) Examine a plankton sample containing live *Daphnia*. Begin by using the naked eye to observe the swimming motion of these small and other small crustaceans. *Daphnia* swim by what appears to be the breaststroke using their large second antennae. WRITE down what you observe about their movement.
2) Examine a prepared slide or a wet-mount slide of *Daphnia*. Live *Daphnia* usually move too quickly for easy viewing under a compound scope but they can be slowed down by adding a viscous material like "Protoslo" to your slide before placing the cover slip or you can put some *Daphnia* in a small container at the beginning of lab and sprinkle some tobacco on the surface of the water. Nicotine diffuses from tobacco into the water and sedates the animals. Observe *Daphnia* under a compound scope and DRAW what you see. Use **Fig. 10.18** to help you identify the anatomy of your specimen.

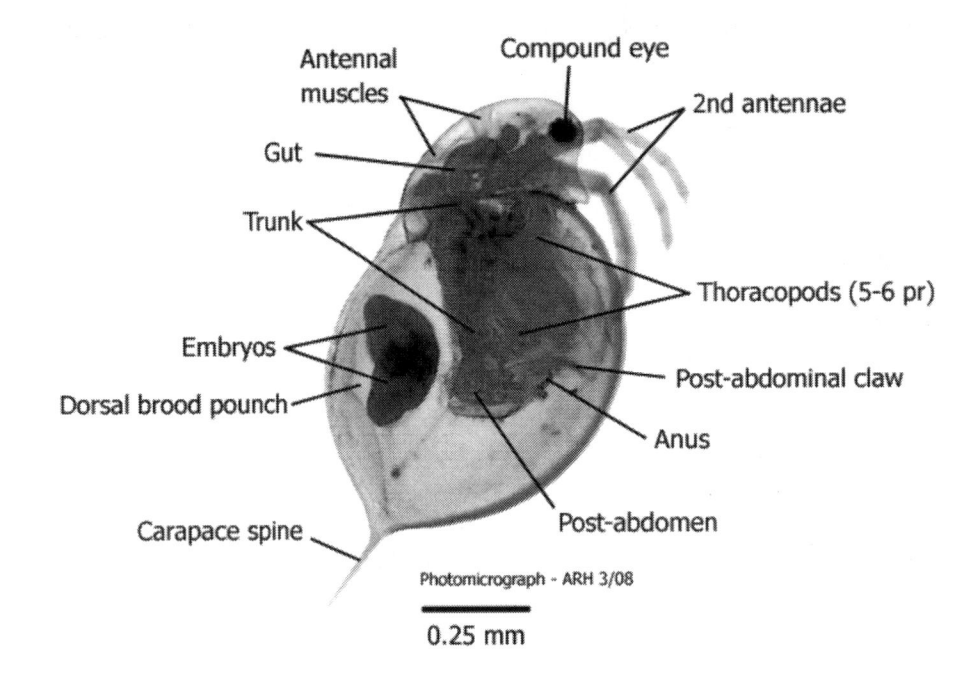

Figure 10.18. Anatomy of the water flea, *Daphnia*, lateral view. (Image: ARH)

Work with a partner as you examine crayfish and the blue crab. One partner will lead the investigation of the crayfish and the other one will lead the investigation of the blue crab.

Tasks – Crayfish dissection

1) Obtain a preserved crayfish. Refer to **Fig. 10.19** to determine if your specimen is female or a male.

Figure 10.19. Anatomy of the crayfish *Cambarus* thorax and abdomen, **ventral view**; male anatomy, **left side**, female anatomy, **right side**. Note the difference between the locations of the female and male gonopores and sizes of the first two pairs of pleopods. (Image: ARH)

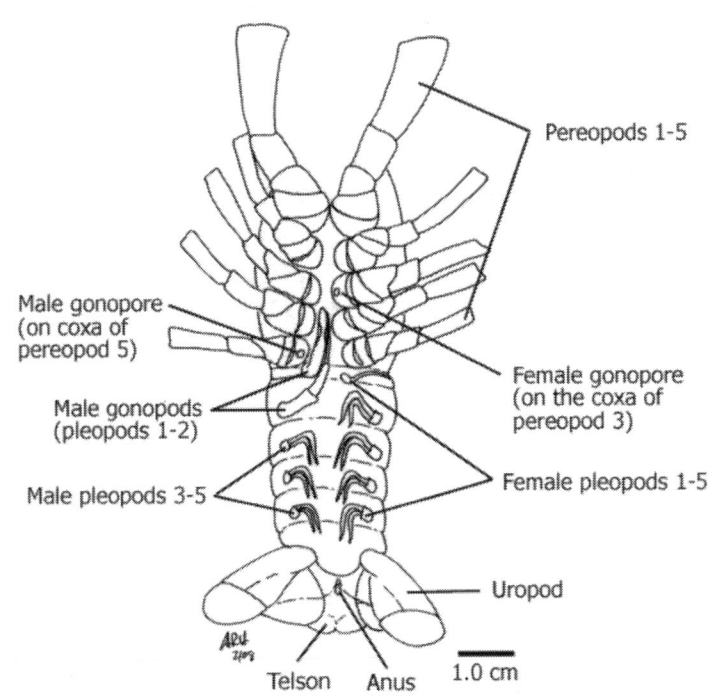

2) Examine the external anatomy of a crayfish. DRAW your specimen and use **Fig. 10.20** to help you identify what you see.

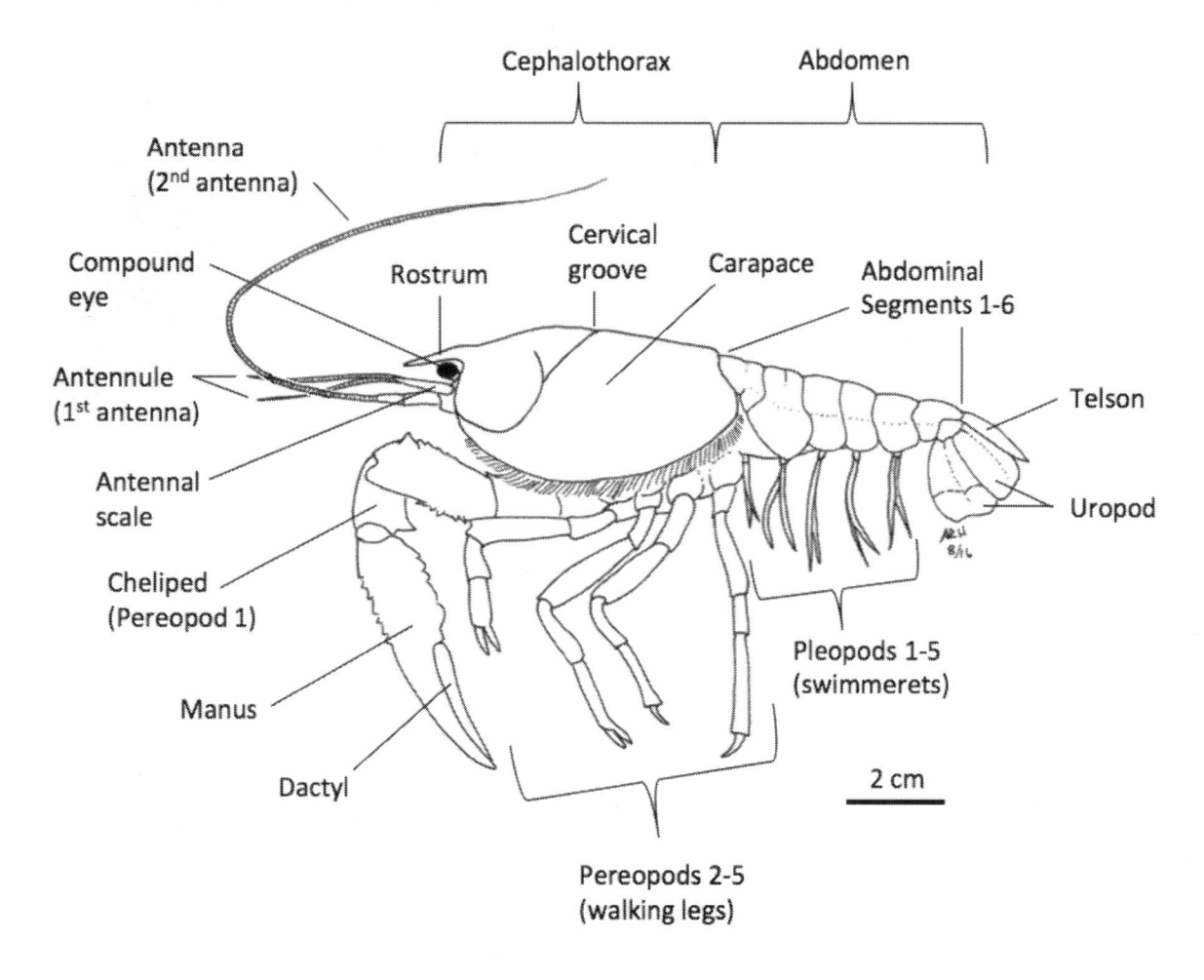

Figure 10.20. External anatomy of female *Cambarus*, lateral view. Pereopods 1-3 are chelate appendages (pincers) while pereopods 3 and 4 are subchelate – appendage tip flexes but does not pinch. (Image: ARH)

3) Remove the carapace. Use fine-tipped scissors to cut the exoskeleton of the carapace starting at the posterior edge of the carapace just off-center of the dorsal mid-line. Continue that cut until you reach the rostrum. Next, make a cut from the leading edge of the rostrum laterally until you reach the ventral edge of the carapace. Insert a probe between the carapace and soft tissues of the body and separate any muscle attachments or other connections between soft tissues and the carapace. Remove the carapace. DRAW your specimen focusing on the gill chamber. Use the top and middle drawings in **Fig. 10.21** to help you identify what you see.
4) Diversity of crayfish appendages. Review **Fig. 10.22** before you remove any appendages. Use a pair of forceps to carefully remove all appendages from one side of your specimen. Remove each entire appendage, including gills that are exopods of some of the appendages. Lay the appendages out in order from anterior to posterior on a paper towel

or in a dissection tray. Refer to **Fig. 10.22** to help you identify each appendage. LEARN the different kinds of appendages and their functions.

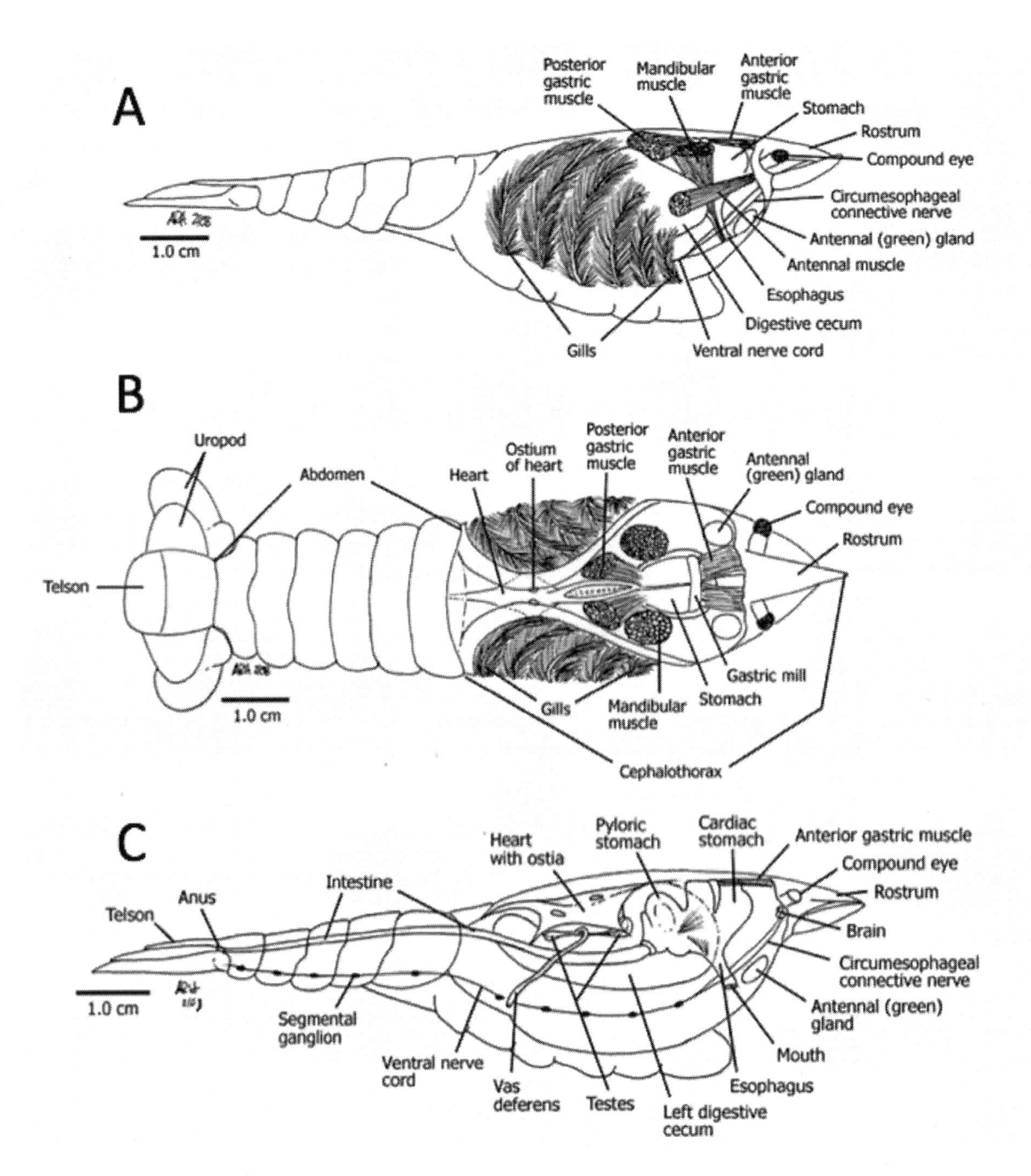

Figure 10.21. Internal anatomy of the crayfish *Cambarus*. Lateral view **(A)** and dorsal view **(B)** of anatomy of the gill chamber and superficial soft tissues of the body cavity with the carapace removed, **C)** lateral view of soft tissues visible when gills and digestive caecum are removed (male specimen) – appendages are not shown. (Image: ARH)

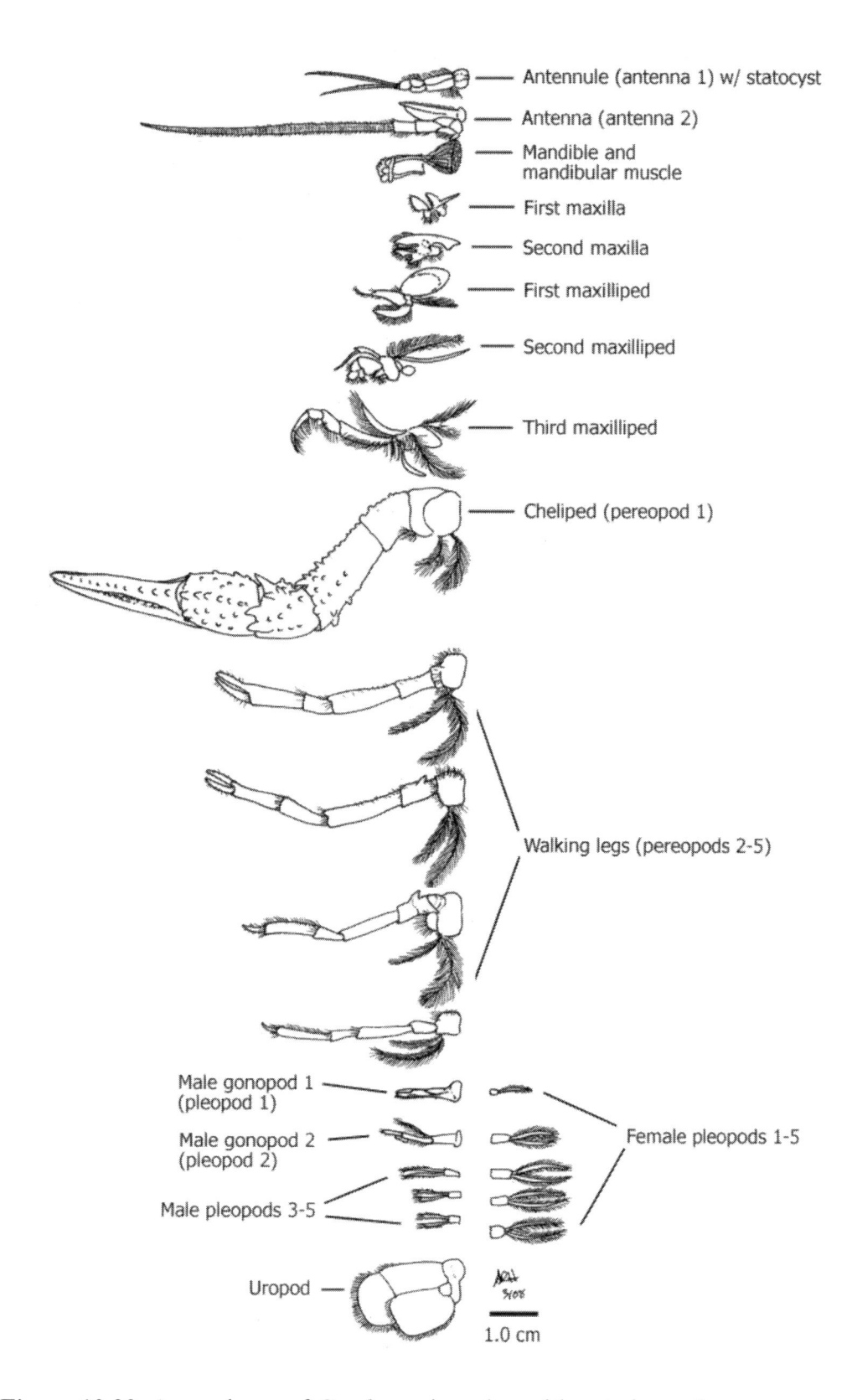

Figure 10.22. Appendages of *Cambarus* in order, with anterior at the top to posterior at the bottom. Note the sexual dimorphism of pleopods. (Image: ARH)

5) Internal anatomy of the crayfish. Remove the thin inner wall of the gill chamber and the large lateral digestive cecum. Look for the long circumesophageal connective nerves passing around the esophagus in the anterior part of the cephalothorax. Look also for the digestive tract and reproductive system. DRAW your specimen. Refer to the bottom drawing in **Fig. 10.21** to help you identify what you see.
6) The intestine. Cut away the exoskeleton from the dorsal half of the abdomen. You should be able to see the intestine running along a dorsal groove in the abdominal muscles. FYI, in recipes for shrimp, crayfish and lobster the intestine is usually referred to as the "sand vein." I guess no one wants to think too much about removing another animal's intestine while they are making dinner.
7) Stomach anatomy. Cut the stomach in half along the dorsal midline. Look for the hardened chitinous teeth on the stomach wall. These hardened structures are the gastric mill. Look for the cardiac and pyloric chambers of the stomach. Refer to the bottom drawing in **Fig. 10.21** for orientation to the structures of the stomach.
8) Ventral nerve cord. Remove the muscles of the abdomen as well as any remaining organs in the cephalothorax. Look for the whitish ventral nerve cord and segmental ganglia that lie along the ventral wall of the exoskeleton. Look also in the rostrum to locate the brain. DRAW your specimen and refer again to **Fig. 10.21**, bottom drawing, for orientation to what you see.

Tasks – Crab dissection

The blue crab *Callinectes* is the main food crab of the East and Gulf coasts of North America. This crab is readily obtained from biological supply companies, though locally available species may certainly be substituted for this exercise.

1) Crab gender. Determine the gender of your crab by looking at the shape of the abdomen. If the abdomen is broad as shown in **Fig.10.23** then your crab is female. If the abdomen is long and narrow it is male.
2) External anatomy. DRAW dorsal and ventral views of your specimen and refer to **Fig. 10.23** to help you identify what you see. Use a probe to lift up the abdomen and expose the pleopods (abdominal appendages, not shown on **Fig. 10.23**). Pleopods of females are long and feather-like; pleopods of males are long, cylindrical and generally lack setae. Male pleopods are specialized for transferring spermatophores to the female and female pleopods are specialized for carrying a mass of eggs called a sponge.
3) Gill chamber and gills. As in crayfish the gills of crabs are external to the body cavity but are housed in gill chamber protected by the exoskeleton. Gills are exopods of several appendages. Open the body cavity by cutting away the dorsal wall of the carapace. Insert the tip of your scissors into the suture line between the posterior edge of the carapace and the first segment of the abdomen. Cut from that point all the way around the dorsal margin of the carapace. Cut as close to the outer edge of the carapace as possible. Once you have completed this cut use a probe to separate any soft tissues from the exoskeleton as you gently lift and remove the dorsal portion of the carapace. If you are not careful you could rip the heart and dorsal wall of the stomach off as you remove the carapace. DRAW all exposed structures and use **Fig. 10.24** to help you identify what you see.

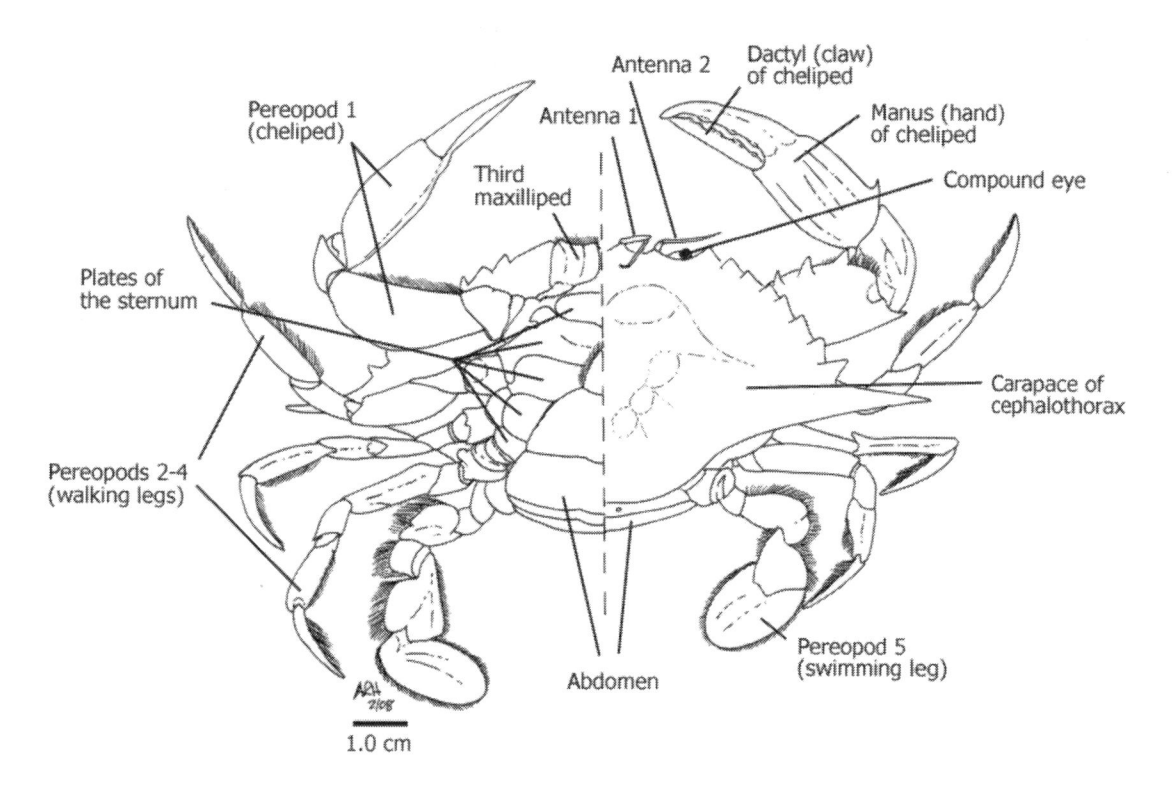

Figure 10.23. External anatomy of female *Callinectes*: ventral view, **left**; dorsal view, **right**. (Image: ARH)

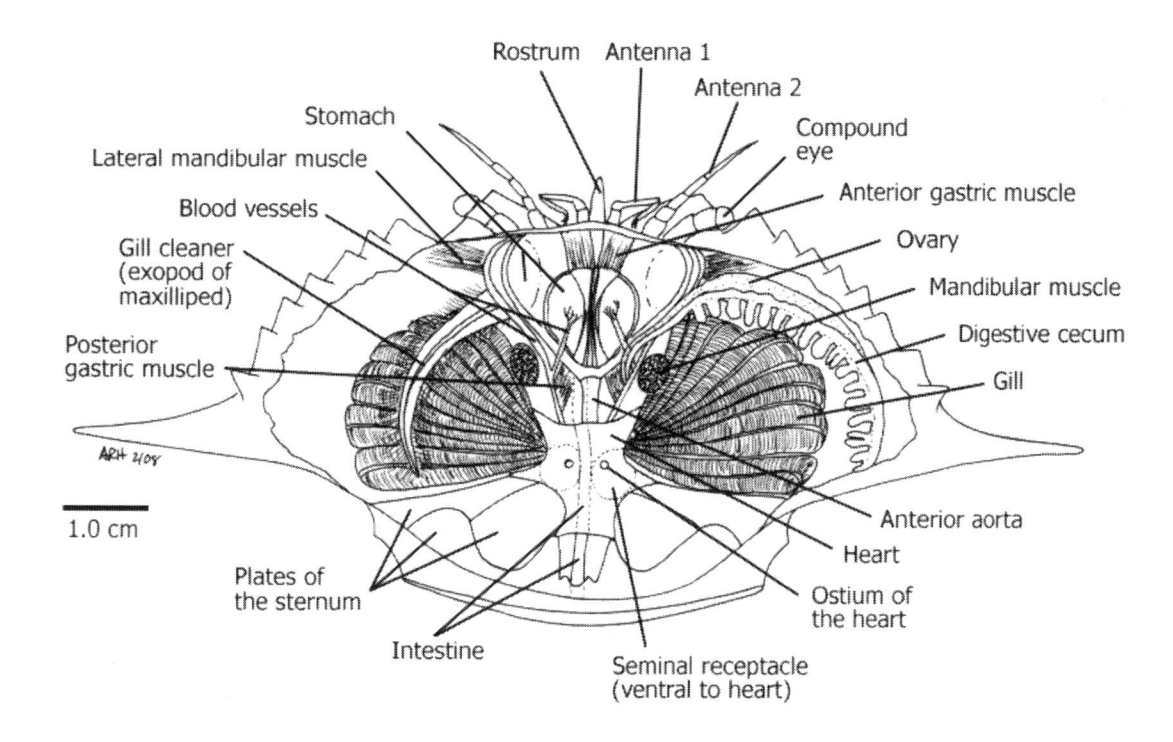

Figure 10.24. Gill chambers and internal anatomy of female *Callinectes*. The digestive cecum, gonads and associated structures are not included on the left side of this figure. (Image: ARH)

4) Remove the digestive cecum and gonads from one side of the body cavity - testes are in the same location as ovaries on **Fig. 10.24**. Removing these structures this makes it easier to observe the gill cleaner. The gill cleaner is attached to the third maxilliped. Once you have found the gill cleaner remove the heart and the portion of the digestive cecum that extends below the heart. You should now be able to see the intestine and if your specimen is a female you will also see a pair of large oblong seminal receptacles.
5) Stomach. Remove the stomach and cut it in half through the dorsal midline. Look inside the stomach for teeth of the gastric mill as well as for the cardiac (anterior section) and pyloric (posterior section) chambers of the stomach; stomachs of crabs and crayfish are similar. DRAW what you see.
6) Appendages. Before you remove any appendages from your crab review **Fig. 10.22** (the order and types of crayfish appendages). Crabs have the same type and number of appendages and in the same order from anterior to posterior as in crayfish though they differ in size and shape. Use a pair of forceps or your hands to carefully remove all of the appendages from one side of your specimen. Be sure that you remove the entire appendage including the gills. This is easier in crayfish than crabs but do your best. Lay the appendages out on a paper towel in order from anterior to posterior. Refer to **Fig. 10.22** again to help you identify each appendage. LEARN the different kinds of appendages, their names, locations and functions.
7) Nervous system. Carefully remove all remaining organs from the body cavity of the crab. Look in the cavity of the rostrum and locate the brain. Next look for the ventral nerve cord and segmental ganglia. Nervous tissue is typically whitish and the nerve cords are thread-like.

<u>Tasks – Crustacean development</u>

1) Developmental stages of crustaceans. Examine prepared slides of all available developmental stages. The nauplius stage is a characteristic life stage produced by all aquatic crustaceans that have swimming larval stages. It is the first larval stage. DRAW all available developmental stages and refer to **Fig. 10.25** to help you identify what you see.

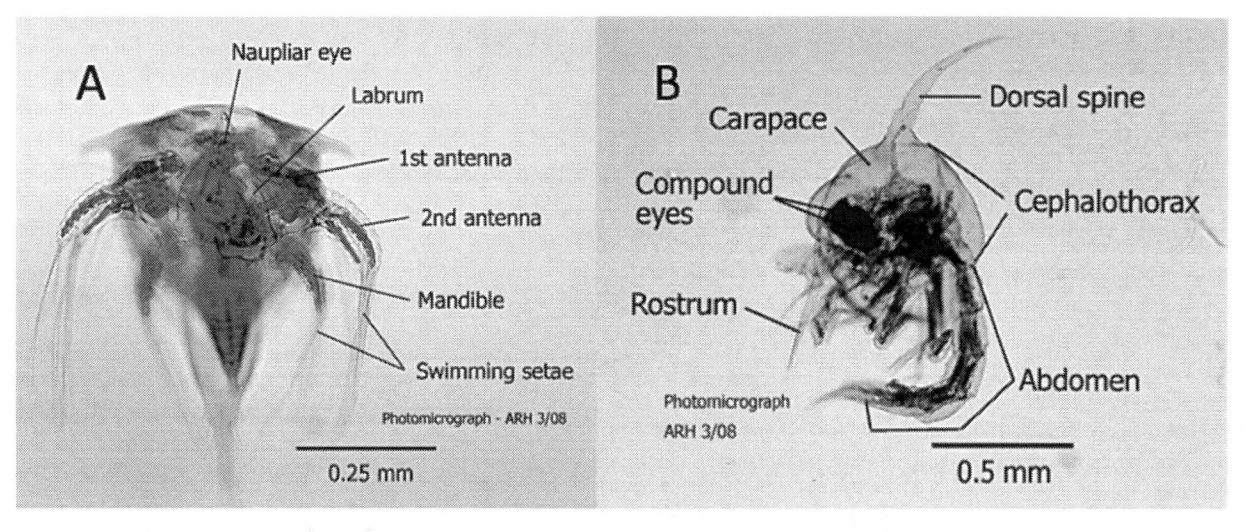

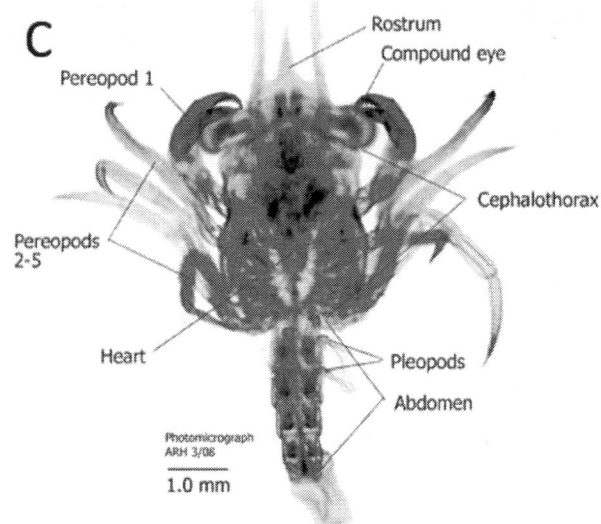

Figure 10.25. Crustacean larval stages: **A)** nauplius, **B)** zoea and **C)** megalops. (Image: ARH)

Subphylum Hexapoda "six feet" – insects and relations

This is the largest subphylum in Phylum Arthropoda. It contains around one million species including the insects and their close kin. Available evidence suggests that hexapods evolved on land. Aquatic forms adapted secondarily to water. Insects play important roles in global ecology. Some insects pollinate huge numbers of plants including crop plants, while other insects are significant crop pests. Hexapods also play important roles as disease vectors, decomposers, scavengers, predators, etc. Characteristics of hexapods are listed in **Table 10.8**.

On a side note, there is currently deep concern about the health of many insect species, especially pollinators such as the honeybee. We rely on bees in particular bees together with other insects to pollinate as much as 80% of our food crops. Unfortunately many domesticated colonies of honeybees are experiencing colony collapse disorder, a condition where entire colonies die without warning or apparent cause. Though explanations are far from conclusive leading theories explaining these deaths include increasing prevalence of bee mites, bee viruses and unintended backlash effects of agricultural pesticides. If losses of these important pollinators continue we could lose an incredibly valuable ecosystem service and some of our crop productivity could be in peril.

Table 10.8. Characteristics of Subphylum Hexapoda (after Brusca, *et al.*, 2016).

- Body with three tagma
    - Head
    - Thorax
    - Abdomen
- One pair of antennae
- Compound eyes and simple ocelli
- Three pairs of thoracic uniramous legs
- All legs with 6 articles
- Gas exchange via spiracles and trachea
- Fused exoskeleton of the head constitutes a supportive structure called the tentorium
- Malpighian tubules for excretion
- Two pairs of thoracic wings (in insects)

Tasks – Insect anatomy – lubber grasshopper

1) External anatomy. Use a magnifying glass to look for spiracles on the abdomen, mouthparts of the head and simple eyes of the head. DRAW your specimen and use **Fig. 10.26** to help you identify what you see.

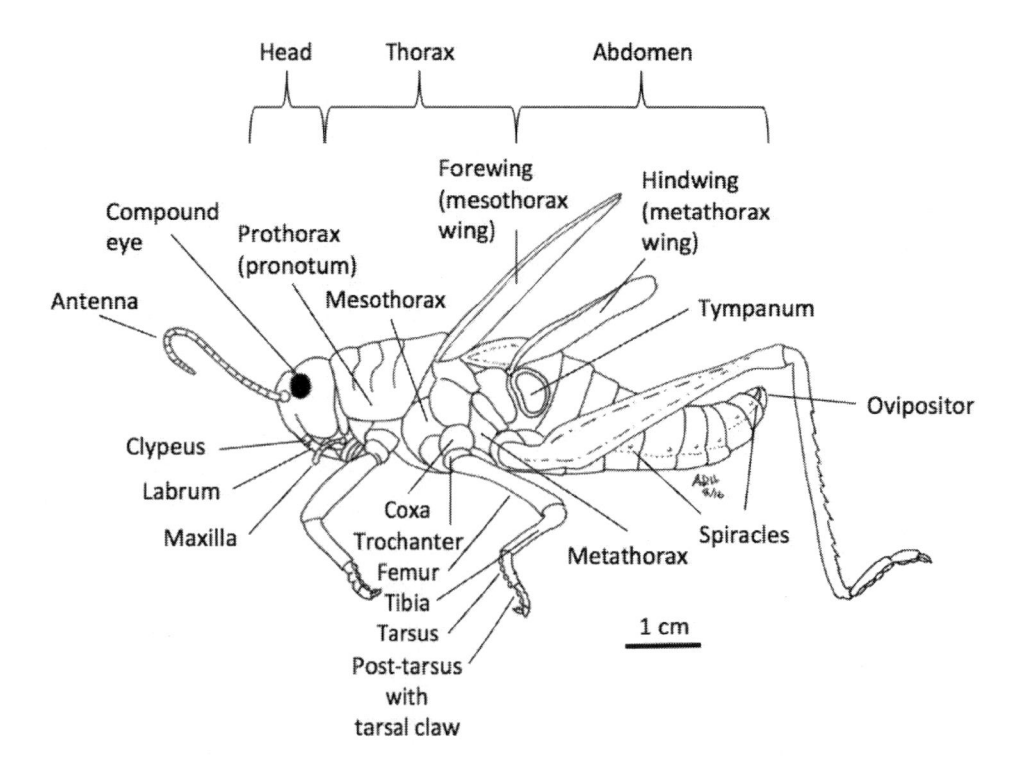

Figure 10.26. Anatomy of the lubber grasshopper *Romalea guttata*. (Image: ARH)

2) Appendages. Use a pair of forceps to remove all of the appendages from one side of the body. Leave the coxa (basal article) of walking legs attached to the body. If you try to remove the coxa with the rest of the leg this will pull muscles inside the body cavity that are attached to the inner wall of the exoskeleton and this will disturb the internal anatomy you have yet to observe. DRAW the leg and LEARN the names of the articles of the 2nd thoracic appendage as shown in **Fig. 10.26**.

3) Internal anatomy. Use scissors to remove the wings and any walking legs that are still attached to the body (leave the coxa of all appendages in place). Use scissors to cut around the edges of the pronotum (see **Fig. 10.26**) and use a probe to scrape the inner surface of the pronotum as you remove it. This separates soft tissues from the pronotum. Use scissors to make a longitudinal cut through the dorsal wall of the exoskeleton that runs the entire length of the thorax and abdomen. Make short lateral cuts along the exoskeleton and use insect pins to attach the exoskeleton to the floor of a wax-bottomed dissection tray. This gives you a dorsal view of the internal anatomy of your specimen. Immerse your specimen. You should see a thin layer of tissue (often reddish) covering much of the internal body cavity. This reddish tissue includes longitudinal and circular abdominal muscles. Look for the diamond shaped heart and dorsal vessel located dorsal to the midgut. After you have made a few observational notes use forceps (not scissors!) to carefully remove the reddish tissue layer. DRAW what you see. Refer to **Fig. 10.27** to help you identify structures of the internal anatomy of your grasshopper.

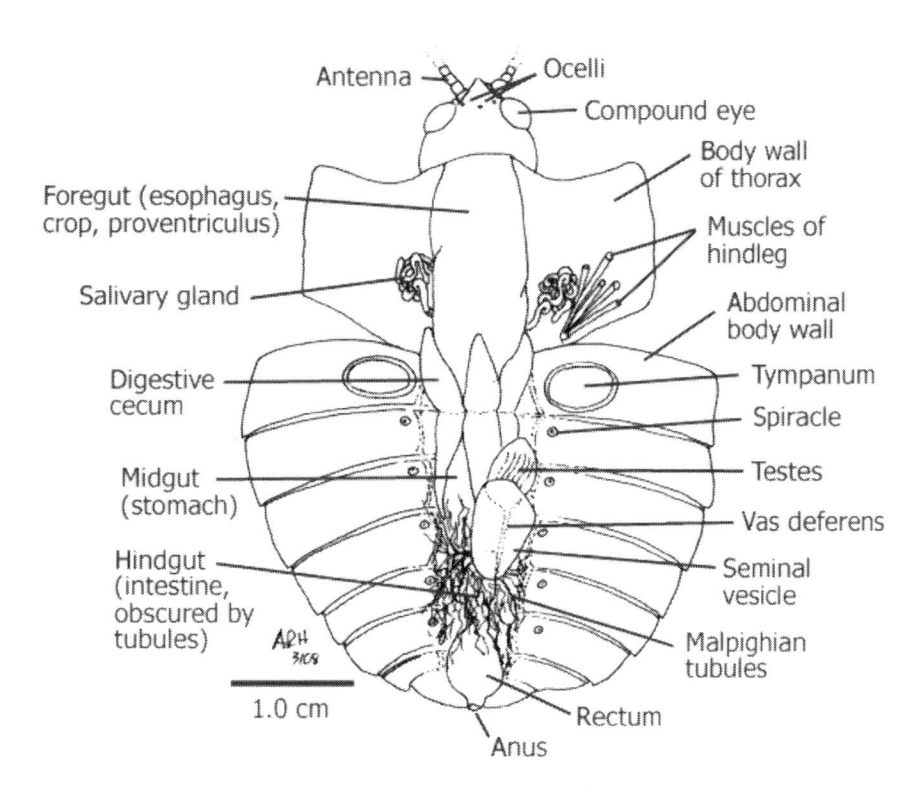

Figure 10.27. Dorsal view of the internal anatomy of a male lubber grasshopper *Romalea guttata* with the dorsal vessel and heart removed. (Image: ARH)

4) The reproductive system. Reproductive organs are paired structures that lie mainly along the dorsal surface of the midgut and hindgut of the digestive tract. Remove one set of the reproductive organs to expose structures ventral to them. DRAW what you see.
5) Nervous system. Remove all of the organs from the body cavity but do not disturb the thin layer of tissue lining the floor of the body cavity. Immerse your specimen and use a magnifying lens or dissection scope to locate the whitish to translucent thread-like structures of the ventral nerve cord, segmental ganglia of the abdomen and many radiating nerves of the thoracic ganglia. DRAW what you see.

Task – Insect diversity

Insects are extremely common and ecologically important so you should become familiar with additional diversity within this group. LEARN the order names and representatives included in **Table 10.9**.

Table 10.9. Selected insect orders and representatives. (Images: ARH)

Insect order name	Representative	Photo of representative
Thysanura "fringe tail"	Silverfish	
Ephemeroptera "temporary wing"	Mayfly	
Odonata "toothed"	Dragonfly	
Plecoptera "braided wing"	Stonefly	

Blattoidea "cockroach form"	Cockroach	
Mantodea "prophet form"	Praying mantis	
Phasmida "phantom"	Walking stick	
Dermaptera "skin wing"	Earwig	
Orthoptera "straight wing"	Grasshopper	
Hemiptera "one-half wing"	True bug	
Psocodea "dust like"	Body louse	

Hymenoptera "membrane wing"	Bee	
Coleoptera "sheath wing"	Beetle	
Siphonaptera 'tube without wings"	Flea	
Diptera "two wings"	Horsefly	
Trichoptera "hair wings"	Caddis fly	
Lepidoptera "scale wings"	Butterfly	

Group questions

1) Describe one specific example of how specialized appendages contributed to the success of an arthropod species.
2) Hexapods are the dominant invertebrate taxon on land but do you think keeps these animals from becoming the dominant group in marine communities as well?
3) Horseshoe crabs spend most of their time on soft sediments where they move slowly along searching for small invertebrates that they dig up, crush and ingest. In what ways is the horseshoe crab body plan well adapted to life on soft substrates?

Clade Panarthropoda – Glossary

Abdomen – posterior body tagma that typically houses the midgut, openings to respiratory structures, gonads and excretory organs

Aculeus (stinger) – barb-like extension of the scorpion telson that penetrates the body wall of other organisms and injects toxins

Anabiosis/cryptobiosis – survival strategy used by tardigrades to survive difficult environmental stress, involves extreme desiccation and entering a state of suspended animation

Antenna – elongate segmented tactile and chemosensory structure (do not confuse antennae with tentacles that are non-segmented sensory structures, as in molluscs)

Antennal (green) gland – excretory organ with an opening located ventral to the antenna and near the excurrent flow of water leaving the gill chamber

Antennal scale – exoskeletal plate located alongside the base of the first antenna (antennule)

Antennule – first antenna of crustaceans, biramous tactile and chemosensory structure

Article – one section of a jointed appendage

Axial furrow – depression marking the division between the axial and pleural lobes in trilobites

Axial lobe – raised middle lobe of thoracic segments of trilobites

Axial ring – exoskeletal covering of an individual segment in the axial lobe

Biramous appendage – jointed appendage with two major branches, the exopod and endopod

Body tubercles – bump-like chemosensory and mechanosensory organs structures that cover the body surface of onychophorans

Book gills – respiratory structure in some chelicerates with many sheet-like gills reminiscent of pages of a book that are protected by a hardened gill cover, e.g., horseshoe crabs

Book lungs – respiratory structures in some chelicerates that are sheet-like in structure and are completely enclosed in chambers accessible only via spiracles, e.g., spiders and scorpions

Capitulum – penetration and attachment organ of parasitic ticks, does not contain the brain

Carapace – shield-like exoskeletal structure that covers all or part of the dorsal surface

Cardiac stomach – anterior portion of the crustacean stomach used for temporary food storage and houses the gastric mill

Cephalic shield – exoskeletal plate that protects the head of centipedes

Cephalon – anterior tagma of trilobites, and is the first body segment of pycnogonids

Cephalothorax – body tagma containing structures of the head and thorax

Cervical groove – groove-like depression in the carapace of the cephalothorax that indicates the separation of the structures of the head and thorax

Chela – an enlarged pinching appendage, may be a modified pedipalp (e.g., scorpions) or walking leg (e.g., crustaceans)

Chelicerae – first pair of appendages in chelicerates

Cheliped – appendage that bears an enlarged chelate or pinching structure, e.g., first pereopod of a crab

Chilarium – small club-shaped appendage located posterior to the last walking leg in horseshoe crabs, has spinous processes like gnathobases of walking legs and marks the posterior end of the food groove

Circumesophageal connective nerve – nerves that pass from the brain around both sides of the esophagus and then rejoin ventral to the esophagus and form the ventral nerve cord

Claw gland – organ in the leg that secretes claws in tardigrades

Cloaca – opening through which more than one organ system releases its product from the body, e.g., digestive system and reproductive system

Clypeus – an exoskeletal plate on the front of an insect head, marks the lower boundary of the head and the labrum is attached to the clypeus ventrally

Collum – first segment of the trunk in millipedes

Compound eye – optic organ made of many ommatidia (individual light collecting structures) that can produce high quality images

Coxa – first article (segment) of jointed appendages, connects an appendage to the body

Cuticle – non-elastic chitinous outer body covering secreted by the epidermis

Dactyl – movable claw of a chelate appendage, word root means "finger"

Deutocerebrum – middle lobe of the three-part brain, usually associated with antennae and olfactory senses

Digestive cecum – an organ that is a lateral outpocketing of the gut where extracellular digestion and food absorption takes place

Diplosegment – segments of the trunk of millipedes that are tagma made of two fused body segments, each diplosegment bears two pairs of legs, two pairs of spiracles, etc.

Dorsal brood pouch – dorsal chamber in *Daphnia* that houses and protects developing embryos

Dorsal shield – exoskeleton plate covering the dorsal body wall of the trilobite thorax

Doublure – edge of the exoskeleton that contains no organs but is two-layers thick and provides extra protection

Ecdysis – process by which groups in the Panarthropoda undergo molting

Epistome – structure that surrounds or is located just anterior to the mouth of a centipede

Esophagus – tube that carries food to the stomach

Exoskeleton – hardened cuticle of arthropods, may be sclerotized (hardened) or calcified

Eyespot – light sensitive cell or cluster of cells, not an ocellus

Facial suture – grooves on the cephalon of some trilobites that are useful in determining lifestyle and taxonomic affiliations

Food groove – depression in the outer body surface where food is manipulated and sometimes partially ground up prior to being ingested, leads to the mouth

Foregut – the cuticle-lined anterior portion of the gut

Forcipule (fang) – appendages of centipedes that are modified to pierce the body wall of prey and inject toxins

Fovea – a depression located in the center of the spider prosoma, is the location where sucking muscles of the stomach attach to the dorsal body wall and is an important characteristic for identifying spiders

Gastric mill – hardened tooth-like structure used to grind up organic matter

Gastric muscle – anterior and posterior gastric muscles manipulate the gastric mill and move food through the stomach

Gena – facial grooves of some trilobites, useful in determining lifestyles and taxonomies

Genital operculum – in horseshoe crabs it is the anterior-most exoskeletal covering of the book gills and also bears the genital opening, in scorpions it is a midventral exoskeletal plate that covers the genital opening

Gill cleaner – in crabs, the feather-like exopod of the third maxilliped that sweeps constantly over the gills and dislodges and removes foreign material

Glabella/glabellular furrow – grooves of the cephalon of some trilobites, useful in determining lifestyle and taxonomic affiliations

Gnathobase – knob-like base of appendages in horseshoe crabs, these bear many spinous processes that can grind food as they move it toward the mouth

Gnathochilarium – large plate-like structure just posterior to mandibles in millipedes, most likely comprised of fused maxillae

Gonopod (male) – pleopods 1 and 2 of crustaceans that are modified in males to transfer spermatophores from the male gonopore to the female gonopore during mating

Gonopore – opening of the reproductive system

Head – tagma that houses the brain, mouth and various sensory organs

Hemocoel – fluid-filled body cavity in open circulatory systems

Hindgut – posterior section of the digestive tract, is lined by cuticle and is where feces are prepared for release

Hypoproct – part of the telson in millipedes, is a shield-like exoskeletal plate located ventral to the anus

Hypostome – a structure located between the chelicerae of ticks that bears backward pointing tooth-like structures that attach the tick to the host body wall

Idiosoma – the body tagma of ticks

Jointed appendage – an appendage covered by exoskeleton and bearing internal muscles that allow the articles of the appendage to flex and extend

Labrum – a flap-like structure of insects that covers the mandibles and produces a partially enclosed chamber where food can be manipulated and digestive enzymes can be extruded to begin digestion and food processing before being swallowed

Lateral process of the body (pedestal) – lateral projections of the trunk of pycnogonids that contain extensions of the intestine and provide points of attachment for walking legs

Lobopodia – non-jointed legs of onychophorans and tardigrades

Malpighian tubule – thin tubular excretory organ that releases uric acid into the digestive tract at the junction of the midgut and hindgut

Mandibles – chewing or slicing mouthparts

Mandibular muscle – large muscle used to manipulate the mandibles

Manus – rigid portion of chelate appendages, word root means "hand"

Maxillae – mouthparts used to taste and manipulate food

Maxilliped – appendage used as a fang in centipedes, used for food manipulation in crustaceans

Median eyes (scorpions) – simple ocelli

Mesosoma – middle tagma of scorpions, contains structures of the digestive, respiratory, excretory and reproductive systems

Mesothorax – middle segment of the insect thorax, bears a pair of walking legs and the anterior pair of thoracic wings

Metasoma – posterior tagma of scorpions, is multisegmented and bears the barbed aculeus

Metathorax – third segment of the insect thorax, bears a pair of walking legs and posterior pair of thoracic wings

Midgut – portion of the digestive tract that is not lined by cuticle

Naupliar eye – anatomically simple light sensory organ comprised of three photosensitive cells surrounded by pigmented cells

Nauplius larva – first post-hatching larval stage of aquatic crustaceans, bears three pairs of appendages

Opisthosoma – posterior body tagma of chelicerates, contains organs of the abdomen

Oral/slime papillae – a pair of knob-shaped structures on the head of onychophorans, used to shoot streams of sticky slime

Ostium of the heart – opening through which hemolymph is taken into the heart

Ovigers – appendages of pycnogonids that bear clusters of eggs

Ovipositer – structure through which female insects release eggs

Paraproct – exoskeletal plates that are part of the telson of millipedes that flank the anus

Pectines – primary chemosensory organs of scorpions, located on the ventral surface of the mesosome

Pedicel – narrow connection between the prosoma and opisthosoma of spiders

Pedipalps – second pair of appendages in chelicerates, can be used for grasping, piercing, signaling, etc.

Pereopods – thoracic appendages of crustaceans, includes one pair of chelae (1st pair) and four pairs of walking legs in crayfish and crabs

Pharyngeal bulb with placoids – muscular organ in tardigrades that houses hardened structures (placoids) that are used to crush food before it enters the midgut

Pleopods – abdominal appendages of crustaceans

Pleural lobes – lateral lobes of thoracic segments in trilobites

Pleurite – lateral extension of the exoskeleton of an individual segment in trilobites, also thickened lateral plate of the exoskeleton in other arthropods

Proboscis – extension of the head that bears the mouth

Pronotum – exoskeletal plate that covers the anterior portion of the thorax in insects

Prosoma – anterior tagma of chelicerates, includes structures of the head and thorax

Prothorax – first segment of the insect thorax, bears a pair of legs but no wings

Protocerebrum – upper lobe of the brain in animals with three-part brains, typically associated with light sensory organs (eyes)

Pusher leg – fourth pair of walking legs in horseshoe crabs, the terminal end bears three leaf-like blades that can be splayed out and provide a broad surface for pushing against soft substrates without sinking into them

Pygydium – posterior tagma of trilobites

Pyloric stomach – posterior portion of the crustacean stomach where inorganic and organic particles are separated, organic particles are moved to a digestive cecum for digestion and inorganic particles are moved into the intestine for elimination from the body

Rostrum – portion of the crustacean carapace that extends farther forward than the eyes

Salivary gland – organ that produces mucus that helps lubricate the movement of food through the gut and sometimes also secretes digestive enzymes

Segmental ganglion – a cluster of nerves on the ventral nerve cord that is associated with one segment of the body

Seminal receptacle – sac-like organ that receives sperm during mating, sperm may be used immediately or stored for future use

Sister taxon – closest related taxon to another taxon

Spinous pad – portion of the leg of onychophorans that contacts the substrate during normal locomotion, provides protection and traction for the lobopod

Spiracle – opening through which air is pulled into trachea or a book lung of terrestrial panarthropods

Sternite – thickened ventral plate of the exoskeleton

Sternum – exoskeletal structures covering the ventral surface of a crab cephalothorax

Stylet – needle-like structure secreted by tardigrades that is used to pierce plant cells, tardigrades usually have two stylets at a time

Stylet gland – organ in tardigrades that secretes the stylet, a pair of these glands are located alongside the mouth

Subesophageal ganglion – first ganglion of the ventral nerve cord in panarthropods

Swimming setae – setae that are spaced so closely together that appendages that bear them function as paddles instead of sieves due to low Reynolds Numbers

Tagma – multiple segments fuse together to produce specialized body parts, e.g., head, thorax or abdomen of insects

Telson – terminal body part extending beyond the last body segment in some arthropods, may be elongate as in horseshoe crabs or shorter as in crayfish, it never carries appendages and is not considered a true body segment

Tentorium – exoskeletal support structure inside the insect head that supports the brain

Tergite – thickened dorsal plate of the exoskeleton

Thoracic wings – wings derived from the dorsal wall of the thorax in insects

Thoracopods – thoracic appendages of cladocerans, used for feeding and moving water to facilitate gas exchange

Thorax – middle body tagma, houses various internal organs including the stomach and heart and bears appendages

Tracheal system – respiratory system of branching tubules that carries air to individual cells

Tritocerebrum – lobe of the three-part brain closest to the mouth, associated with the second antenna, mouthparts and anterior digestive system

Tun stage – desiccated condition of tardigrades in anabiosis

Tympanum – auditory (hearing) organ in insects

Uniramous appendage – jointed appendage that has no exopod/endopod division

Uropod – flattened appendages attached to the last abdominal appendage in crayfish, lobster, shrimp, etc., used for rapid backward locomotion

Vas deferens – tube that carries sperm from the testes to the male gonopore

Ventral nerve cord – solid cord of nervous tissue running along the ventral body wall, is part of the central nervous system and bears segmental ganglia

Chapter 11: Clade Deuterostomia, Phylum Echinodermata

For many years taxonomists divided bilateral animals into two large taxa based on their embryonic development. Animals that exhibited spiral cleavage and a blastopore that gave rise to the mouth were called protostomes "first mouth." Animals that had radial cleavage and a blastopore did not give rise to the mouth were called deuterostomes "second mouth." Phylogenetic analysis rather surprisingly supports this basic taxonomy but the division is now based primarily on molecular rather than developmental traits. Major clades within the Protostomia include the Spiralia and Ecdysozoa, taxa that were covered in earlier chapters. The Deuterostomia includes three phyla: Echinodermata, Hemichordata and Chordata. These are covered in this and the following chapter.

The focus of this chapter is Phylum Echinodermata, a strictly marine taxon of about 7,000 species assigned to three subphyla and five classes. Characteristics of Phylum Echinodermata are listed in **Table 11.1.**

Table 11.1. Characteristics of Phylum Echinodermata (after Brusca, *et al*. 2016).

- Mesodermally derived endoskeleton comprised of calcareous plates and ossicles
- Larvae with primary bilateral symmetry and adults with secondary pentaradial symmetry
- Deuterostomous development – radial cleavage, enterocoely, mouth not derived from the blastopore
- Water vascular system
- Mutable collagenous connective tissue
- Decentralized central nervous system

This video segment from the *Shape of Life* series introduces the body plan and diversity of echinoderms: http://shapeoflife.org/video/echinoderms-ultimate-animal

Phylum Echinodermata "spiny skin" – Taxonomy

Subphylum Crinozoa "lily animal" and Class Crinoidea "lily like" (not covered in this exercise) – crinoids, feather stars and sea lilies, just over 600 living species. The oral surface faces away from the substrate and bears a centrally located upward facing mouth surrounded by arms that extend outward from the central disc. A stalk supporting the body, when present, is attached to the aboral surface. Arms bear feather-like pinnules that are used for suspension feeding.

Subphylum Asterozoa "star animal" – seastars, brittle stars and basket stars. The mouth faces the substratum and this surface is referred to as the oral surface; five or more arms surround the mouth.

Class Asteroidea "star like" (covered in this exercise) – seastars, about 2000 species. Seastars have a star-shaped (stellate) body with a central mouth surrounded by five or more unbranched arms. The oral surface faces the substrate and bears the mouth and ambulacral grooves. There is no clear demarcation between the arms and central disc in these animals. External podia of the tube feet have internal ampullae. Podia may or may not have suckers. The madreporite is on the aboral surface. The anus is often reduced and is also on the aboral surface, when present.

Class Ophiuroidea "snake like" (covered in this exercise) – brittle stars and basket stars, about 2000 species. This class name is particularly apt and you will see why if you have a chance to observe live brittle stars. These cryptic animals keep most of the body hidden and extend only an arm or two into the water. These animals have stellate bodies but the arms may be unbranched (brittle stars) or branched (basket stars) and the boundary between the arms and central disc is clearly visible. Skeletal plates cover the ambulacral groove. Podia in this class are peg-shaped and lack ampullae. The madreporite is on the oral surface and there is no anus. Arms of these animals readily break off when they are disturbed, thus giving brittle stars their common name.

Subphylum Echinozoa "spiny animal" – sea urchins, sand dollars, sea cucumbers and relations. These animals lack arms.

Class Echinoidea "spine like" (covered in this exercise) – sea urchins, sand dollars and relations. This group of about 1000 species has calcareous plates that are fused together to produce a rigid endoskeleton called a test. Moveable spines cover the body and they use podia for locomotion, respiration or both. Some of these animals also have a complex chewing organ called Aristotle's Lantern.

Class Holothuroidea "sea cucumber" (covered in this exercise) – sea cucumbers, about 1700 species. This includes animals with large worm-like bodies, a mouth is located at one end and the anus at the other. Calcareous ossicles are embedded in the fleshy body wall, and they lack a rigid test. Pentaradial symmetry is expressed in the tentacles surrounding the mouth and the central axis is elongated with the mouth at one end and the anus at the other.

Class Asteroidea

Seastars are important members of marine environments where they are often top predators. Seastars have a global distribution and are members of intertidal and benthic communities. Most people are familiar with seastars because they are often colorful and are easy to spot during low tide and while swimming or snorkeling. Plus, seastars are among the most iconic of marine invertebrate animals.

Tasks - Class Asteroidea

1) External anatomy of the seastar – aboral surface. Look at the aboral surface of a seastar. Locate the madreporite, the small button-shaped structure located on the aboral surface near the junction of two of the arms. These two arms are referred to as the bivium and the other three arms are the trivium. DRAW the aboral surface of your specimen.
2) External anatomy of the seastar – oral surface. Look at the oral surface of your seastar. Identify the mouth and oral spines, the five arms and ambulacral grooves that run along the oral surface of each arm. Look carefully at an ambulacral groove and identify spines and podia - the portion of tube feet that extend beyond the body wall. DRAW what you see.

3) Structures of the body wall. Use a pair of scissors to cut a 1 cm x 1 cm piece from the aboral body wall of one arm of your specimen. Immerse this tissue sample and use a dissection scope to examine it. DRAW what you see. Use **Fig. 11.1** to help you identify structures of the aboral body wall.

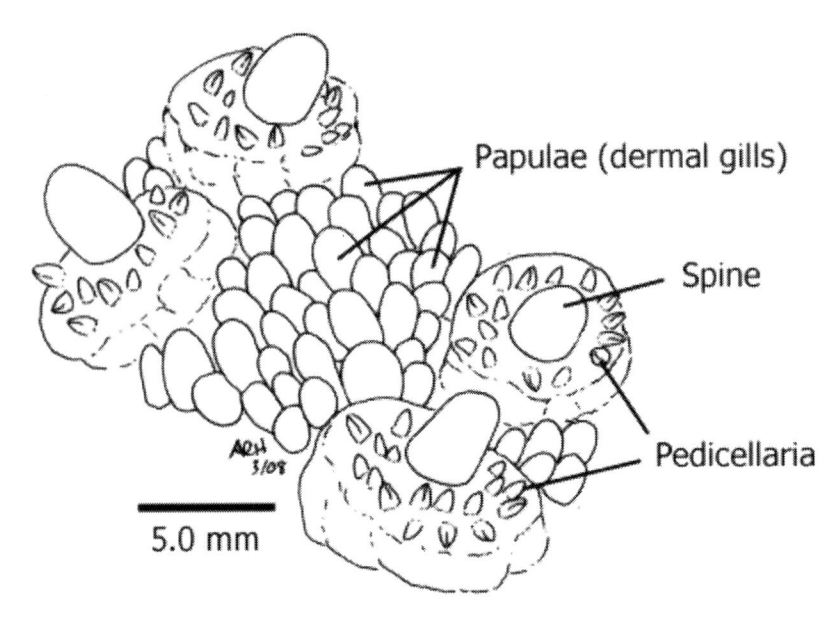

Figure 11.1. Structures of the body wall of the seastar *Asterias*. (Image: ARH)

4) Use a pair of fine-tipped forceps to remove some pedicellaria from near the base of a spine of the body wall. Make a wet mount slide of those pedicellaria and use a compound scope to examine the anatomy of these structures. DRAW a pedicellarium. Work with a partner to deduce how seastar pedicellaria open and close (see **Fig. 11.2).**

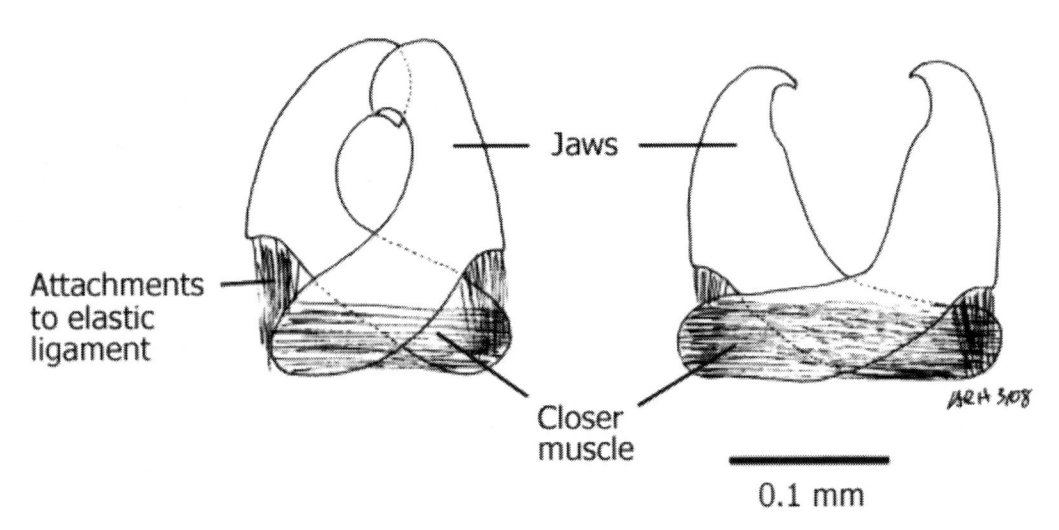

Figure 11.2. Seastar pedicellaria: closed, **left**; and gaping, **right**. (Image: ARH)

5) Internal anatomy. Use a pair of scissors to make a shallow cut around the edges of the central disc, but cut around the madreporite thus leaving it intact and in place. Use a

probe to separate any soft tissues from the body wall as you carefully lift the body wall off of the central disc. Next, cut away the aboral body wall from three arms of the trivium of your specimen. Identify structures of the digestive system in the central disc as well as organs housed in the arms. DRAW the digestive system of a seastar. Refer to **Fig. 11.3** to help you identify these structures.

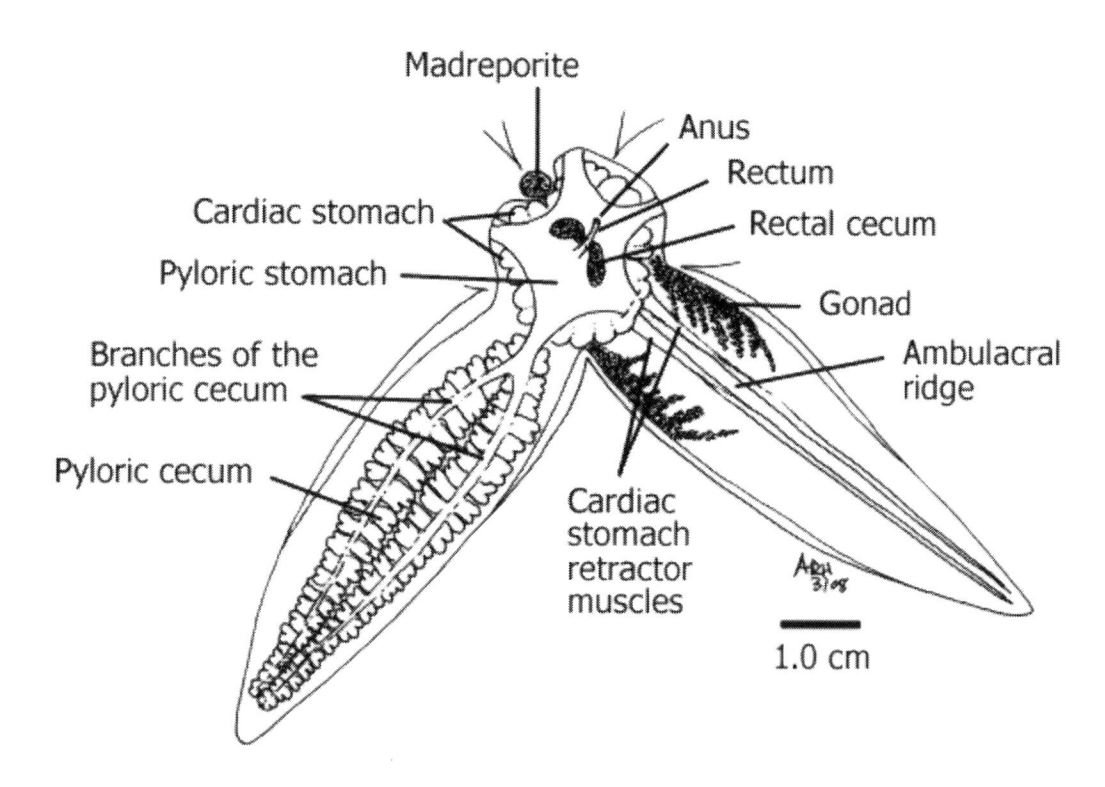

Figure 11.3. Internal anatomy of the central disc and arms of a seastar, aboral view. (Image: ARH)

6) Remove the pyloric cecum from one of the arms of your specimen. Locate the pair of gonads that are situated beneath the pyloric cecum. Identify the cardiac stomach retractor muscles by pulling gently on the cardiac stomach and looking for the nearly transparent cardiac stomach retractor muscles that are attached to the ambulacral ridge. Discuss what these muscles do. DRAW an arm of your seastar showing the location of the gonads and cardiac stomach retractor muscles (see **Fig. 11.3**).
7) Water vascular system. Remove the pyloric and cardiac stomachs from the central disc after you have identified them. Be sure to keep the madreporite and stone canal in place and intact. DRAW the entire water vascular system. Refer to **Fig. 11.4** to help you identify the anatomy of this system.
8) Examine a prepared cross section slide of the arm of a sea star. DRAW what you see and look at **Fig. 11.5** to help you identify structures in that cross-section view.

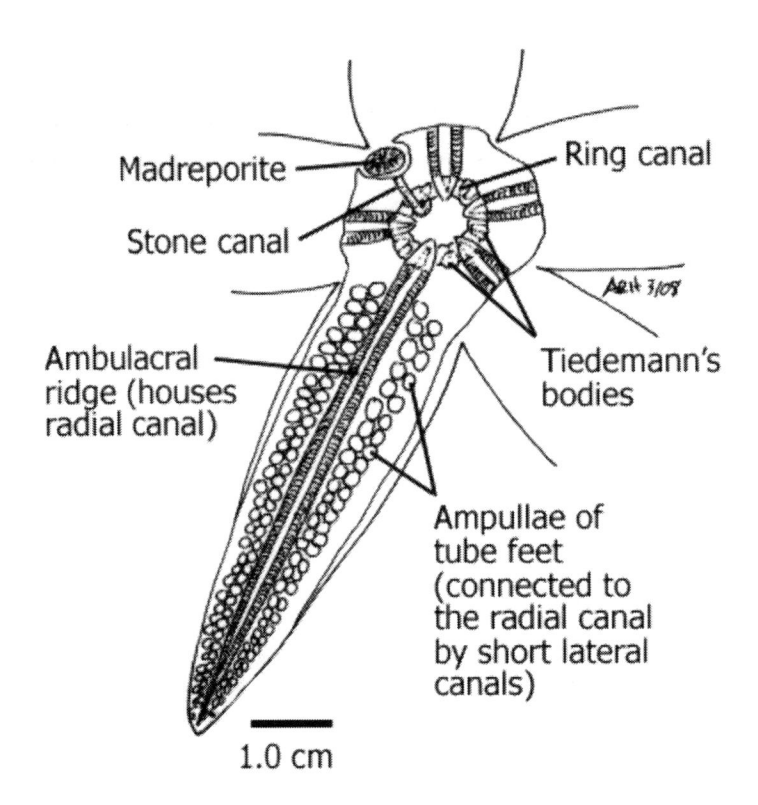

Figure 11.4. The water vascular system of a seastar, aboral view. (Image: ARH)

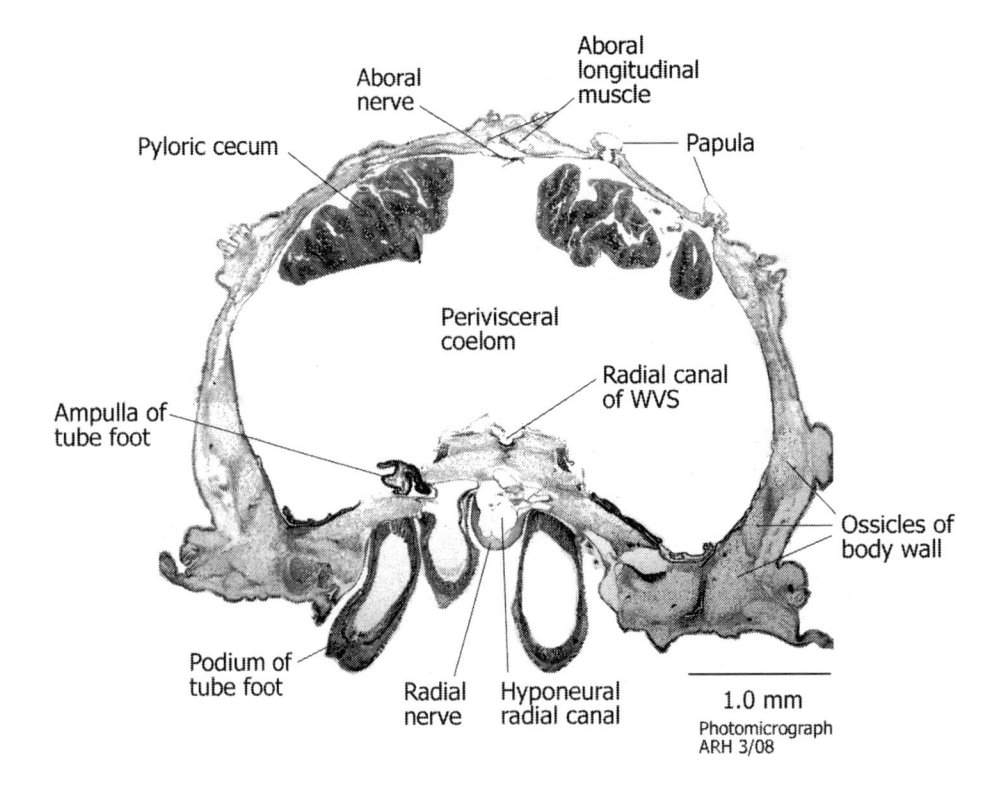

Figure 11.5. Cross-section of the arm of a seastar. (Image: ARH)

Class Ophiuroidea

This class includes brittle stars and basket stars. These animals can be extremely abundant in some environments though they tend to be cryptic. They make a living as suspension feeders, predators, deposit feeders or opportunistic scavengers.

Brittle stars tend to stay largely hidden amongst rocks and rubble, often extending only one or two arms to feed. While most brittle stars have arm spreads of less than 10 cm the largest ones can be 60 cm across. Brittle stars have peg-shaped podia that they use for traction and their main means of locomotion is via muscular action of arms.

Basket stars stay hidden during the daytime and feed at night. When they emerge basket stars crawl up onto prominent locations with good water flow, uncurl their extensively branched arms and carry out suspension feeding and active predation and capture prey up to a few cm long directly from the water column. Basket stars have arm spreads of up to 70 cm across.

Tasks – Class Ophiuroidea, brittle star

1) External anatomy. Observe the oral and aboral surfaces of a preserved brittle star. Before you handle a brittle star be advised that they are aptly named. The arms will readily break off of the rest of the body if you are not extremely careful. Note that the central disc and arms are easily differentiated, something you did not see in the seastar body plan. The arms of brittle stars are quite different than those of seastars. Calcified dorsal and ventral plates and rows of lateral spines protect the arms. These animals have a mouth but no anus. Look for bursal slits on the aboral surface of the central disc. Bursae are invaginations of the body wall used for gas exchange and brooding offspring. DRAW what you see. Refer to **Fig. 11.6** to help you identify the anatomy of a brittle star.

Class Echinoidea

Sea urchins, sand dollars and their close relations are echinoids. Sea urchins play an important role in nutrient cycling and community structure of subtidal communities. Most urchins are opportunistic herbivores and detritovores. When urchin population sizes are kept small by predation or disease drift kelp meets their nutritional needs, but when urchin populations get too large they can consume all available drift kelp and then eat live kelp. This can decimate kelp communities including kelp forests and produce kelp-free habitats called urchin barrens.

Echinoids have rigid endoskeletons called tests that are made of hexagonal $CaCO_3$ plates that are fused together. Additional $CaCO_3$ is added to the outer edges of each plate as an echinoid grows.

Urchins are not inherently dangerous but they are connected with more human injuries than just about any other kind of marine animal. How? Most humans are sufficiently unaware of their environment that they frequently sit on, step on, kneel on or put a hand down on an urchin without looking. When they do this urchin spine penetrate the skin and often break off. These wounds can fester and become serious if not treated.

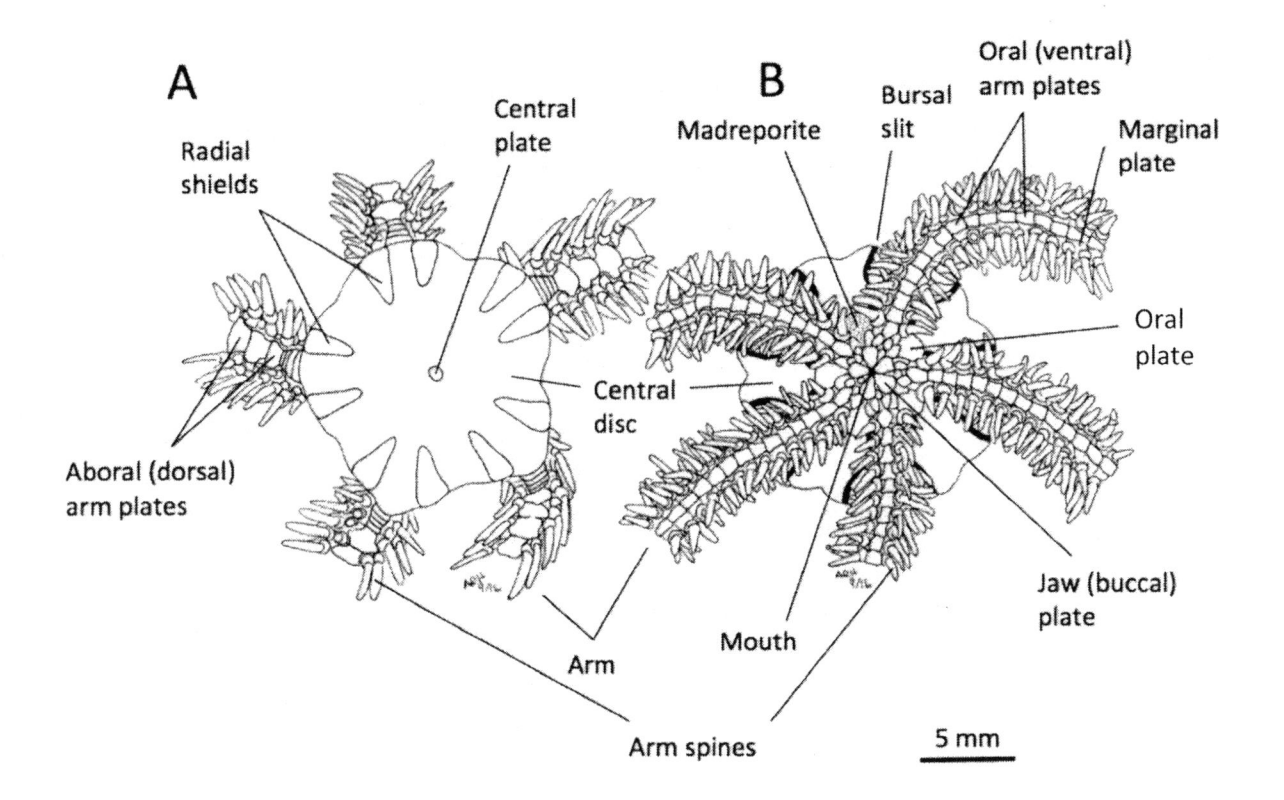

Figure 11.6. Anatomy of a brittle star: **A)** aboral surface, **B)** oral surface. (Images: ARH)

<u>Tasks – Class Echinoidea, sea urchin</u>

Sea urchin

1) Observe a live sea urchin if available. Put it in a large glass bowl filled with seawater or Instant Ocean™. Observe how it uses its tube feet to move. Flip it upside down and observe how it rights itself. Touch one of its spines and see how it reacts to physical contact. RECORD your observations.

2) Use a magnifying glass to examine the oral and aboral poles of a preserved sea urchin. Structures of the aboral end include the periproct, the madreporite, four genital plates, five ocular plates and five gonopores. These are sometimes difficult to identify. Refer to **Fig. 11.7A** to help you identify structures of the periproct that you can see. Examine the oral pole of the specimen. Identify the five teeth and peristomial membrane surrounding the mouth. DRAW the oral and aboral poles.

3) Urchin test. Examine a cleaned urchin test. The test is made of many hexagonal $CaCO_3$ ossicles that are fused together. Some ossicles bear small holes that pass through them while other plates do not. These holes allow podia to extend through the test and remain connected to the rest of the water vascular system. Hole-bearing ossicles alternate with non-hole bearing ossicles are called ambulacral plates and make up a row of plates called an ambulacrum. There are five sets of hole-bearing ossicles and five sets of non-hole bearing ossicles. Non-hole bearing ossicles are interambulacral plates and make up the interambulacrum. DRAW a set of ambulacral and interambulacral plates. Refer to **Fig. 11.7B** to identify these structures.

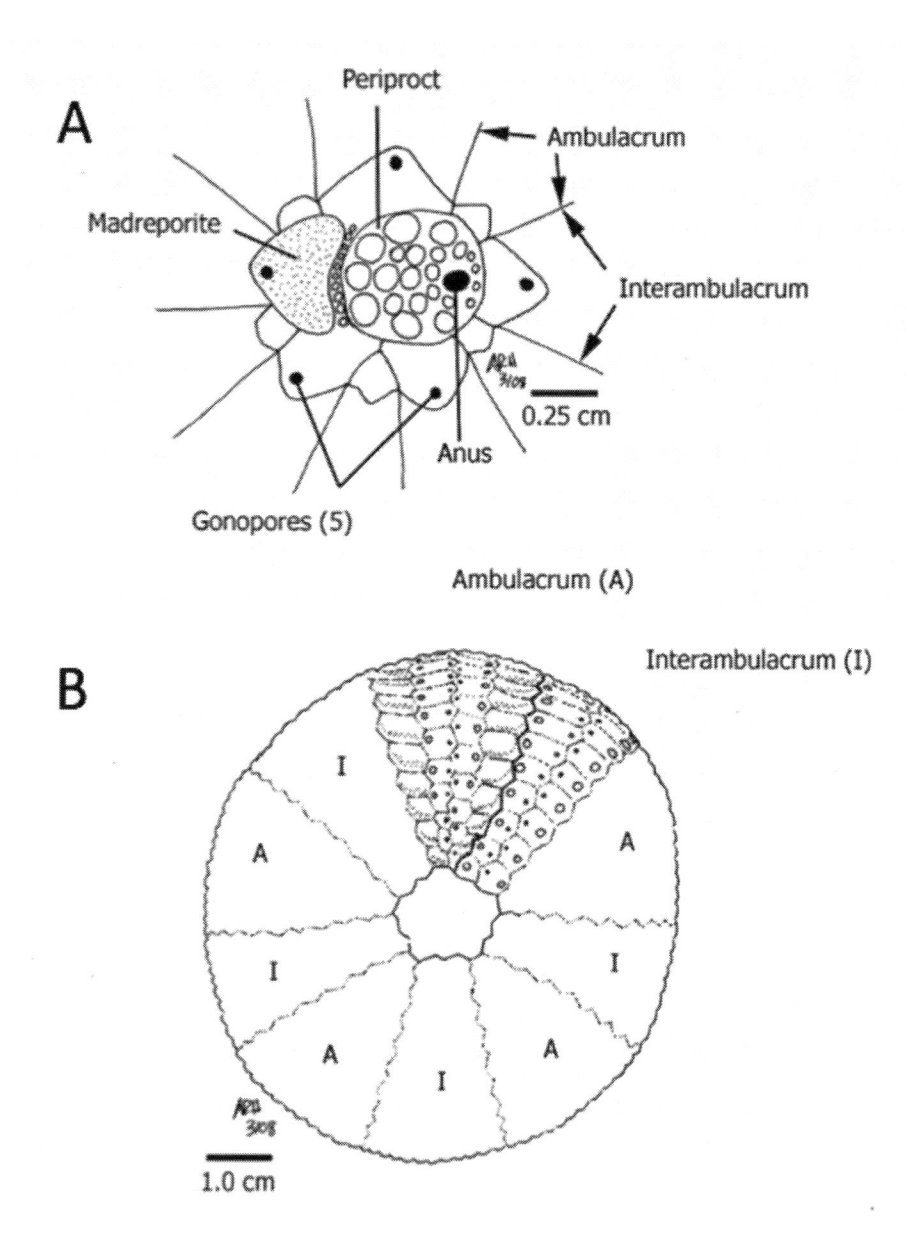

Figure 11.7. External anatomy of the sea urchin *Strongylocentrotus*: **A)** detail of the aboral periproct region, **B)** aboral view of ambulacral and interambulacral plates of the test. (Images: ARH)

4) Use a pair of scissors to penetrate the test anywhere on the aboral half of the test. Remove a small (1 cm x 1 cm) piece of the test and associated structures of the body wall. Immerse your specimen and set it aside. Immerse the small piece of the test and use a dissection scope to observe structures on the surface of the test. Be sure to look for a spine-muscle-tubercle assemblage and pedicellaria. DRAW a spine-tubercle-muscle assemblage.

5) Pedicellaria. Pedicellaria are scattered across the surface of the test between the spines. Use a pair of fine-tipped forceps to remove several pedicellaria from the small piece of the test/body wall that you removed earlier. Make a wet-mount slide of them and use a dissection or compound scope to study their anatomy. DRAW a pedicellarium from your specimen and refer to **Fig. 11.8** to help you identify the anatomy of this structure.

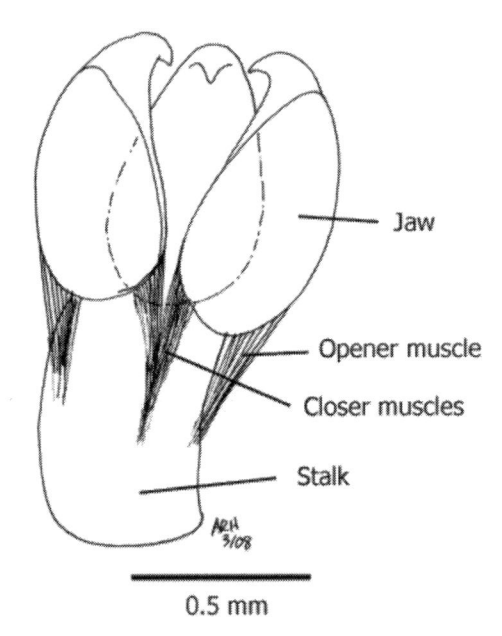

Figure 11.8. Sea urchin pedicellaria have three jaws with closer muscles attached to the inner margin of each jaw and opener muscles attached to the outer edge of each jaw. (Image: ARH)

6) Open the body cavity. Use a pair of scissors to cut all the way around the test. Make your cut above the equator of the test. Hold the oral and aboral halves of the test together until you complete the cut. Use a probe to separate mesenteries that attach the gonads and structures of the digestive system from the inner wall of the aboral half of the test as you lift the aboral half of the test. All soft tissues of the internal anatomy should now be sitting in the oral half of the test. Gently rinse your specimen, change the water in your bowl or tray and re-immerse your specimen.
7) Reproductive system. An aboral view of intact soft tissues of your specimen will reveal five gonads. These can fill much of the space inside the aboral half of the perivisceral coelom, depending on gender and reproductive state of your specimen. Gonads are located along the inner surfaces of the interambulacral plates. Each gonad has its own gonopore in the aboral plate. DRAW what you can see of the reproductive system. Refer to **Fig. 11.9A** to help you identify what you see.
8) Digestive system. Carefully remove the gonads and rinse your specimen again. Most of the digestive tract is located in the oral half of the perivisceral coelom. The digestive tract is held in place by mesenteries that connect it to the inner wall of the test near the equator. DRAW the digestive tract. Look at **Fig. 11.9B** to help you identify what you see.

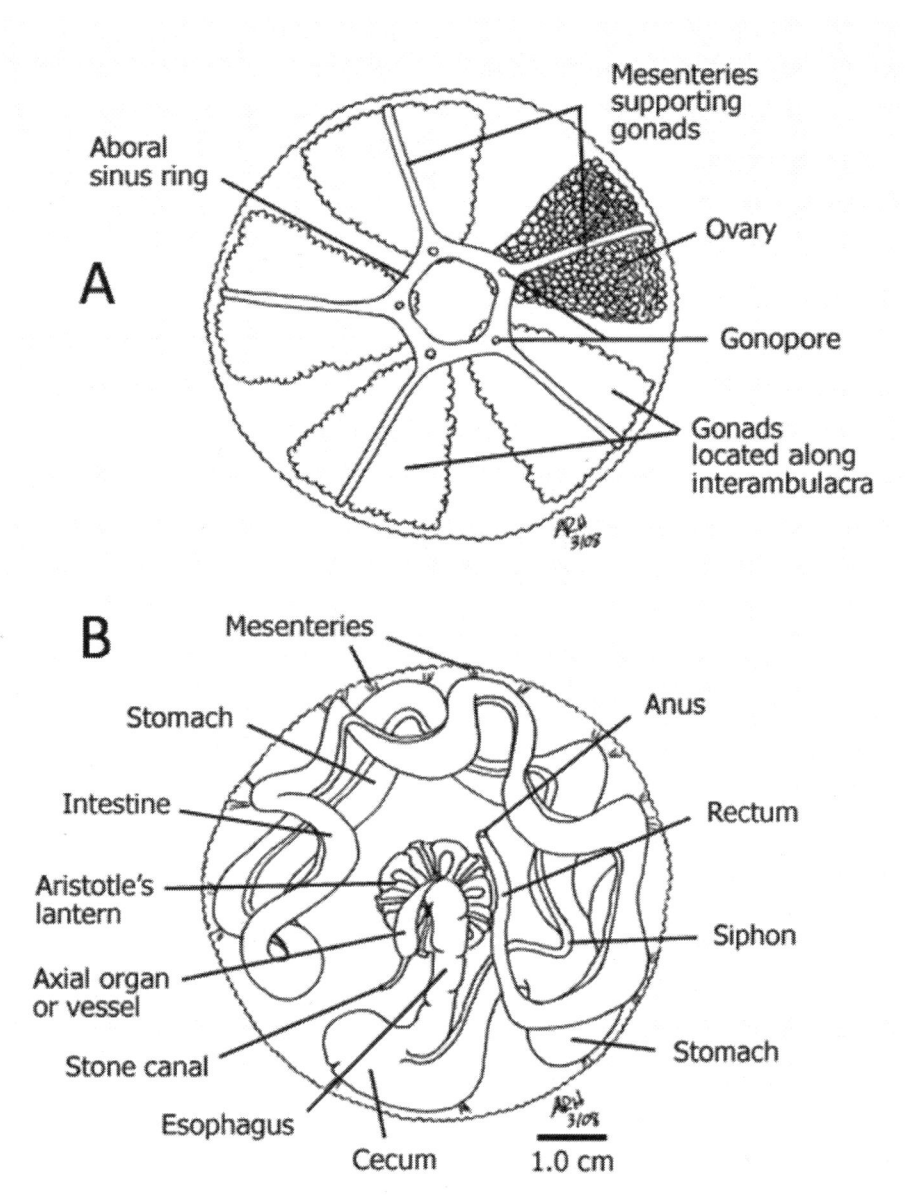

Figure 11.9. Internal anatomy of the sea urchin *Strongylocentrotus*: **A)** aboral view of the reproductive system, **B)** removing the reproductive system reveals the digestive system. (Images: ARH)

9) Water vascular system. The parts of the water vascular system are more difficult to see in urchins than in sea stars. You can however easily see rows of ampullae lining the inner surfaces of ambulacral plates. The ring canal and radial canals can sometimes be seen at the top of the Aristotle's lantern. You should at least be able to see part of the stone canal attached to the axial organ.
10) Aristotle's Lantern. Remove the gonads and the digestive tract and break away the rest of the test except for the part that supports the Aristotle's Lantern. Rinse and immerse Aristotle's Lantern. Examine it using a dissection scope or magnifying lens to locate the structures indicated on **Fig. 11.10** and to see how they interact to move the teeth. Note that there are five sets of everything. Experiment by tugging gently on different parts of

the Aristotle's lantern to see how they move relative to each other. Feel free to disassemble the Aristotle's lantern as you examine it. Enter comments in your lab notebook about what you saw, and about how you think it works. You are not required to draw Aristotle's Lantern, unless you feel compelled to but it is amazing.

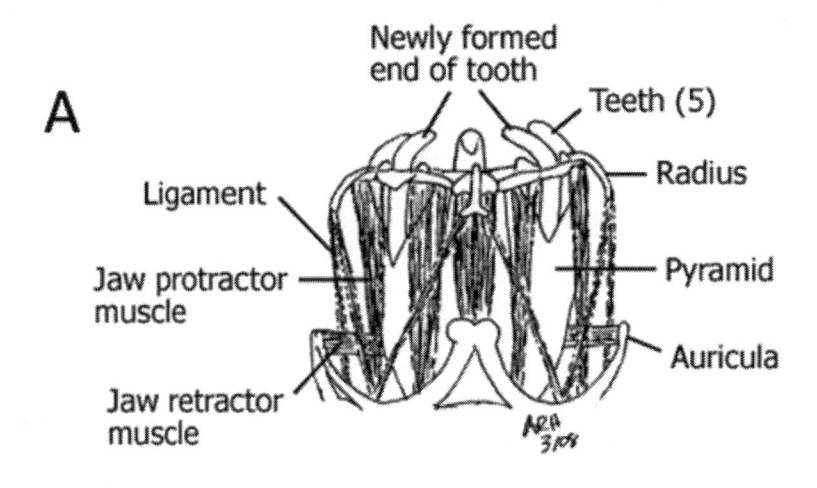

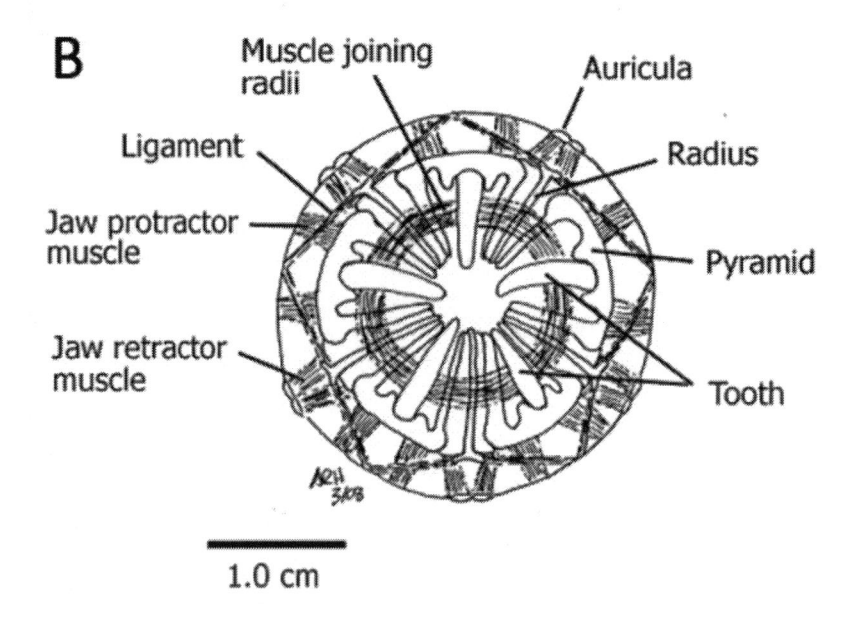

Figure 11.10. Aristotle's lantern: **A)** lateral view, **B)** aboral view. (Images: ARH)

Sand dollars

Sand dollars have bodies that are flattened in the oral-aboral plane. Short spines cover the entire body of these animals. Urchins make a living by suspension feeding. They do this cooperatively. When a current flows over a bed of sand dollars they reorient their bodies so that one edge of the body is embedded in the sand and the other edge sticks up into the flowing water. This produces eddies behind urchins where water flow rates slow and small particles can be captured and moved along food grooves to the mouth.

Tasks – Class Echinoidea, sand dollar

1) Obtain a preserved specimen and study its external anatomy. Compare and contrast the spines of sand dollars with those of sea urchins. WRITE your observations in your laboratory notebook.
2) Obtain a cleaned sand dollar test. Study the oral and aboral surfaces of the test and DRAW what you see. Refer to **Fig. 11.11** to help you identify anatomy of the sea urchin test.

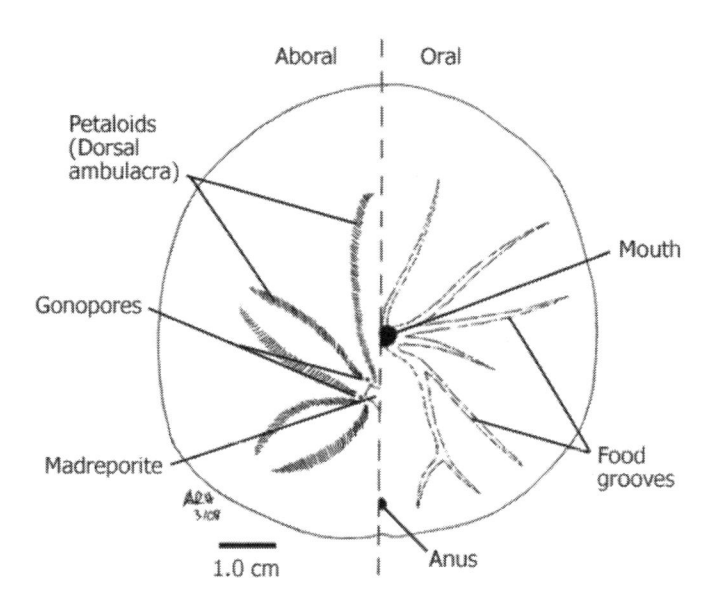

Figure 11.11. External anatomy of the test of the west coast sand dollar *Dendraster excentricus*: aboral surface, **left**, oral surface, **right**. (Image: ARH)

Class Holothuroidea

Sea cucumbers present another variation on the echinoderm body plan. These animals have pentaradial symmetry but in a longitudinal axis. The mouth is located at one end and the anus at the other end of the worm-like body. Sea cucumbers make a living by using their tentacles for suspension feeding or deposit feeding. If you have a chance to observe living sea cucumbers you will see them periodically stick their tentacles into their mouth so can lick off particles that they captured or collected.

Tasks – Class Holothuroidea

1) External anatomy. Obtain a preserved sea cucumber. Tentacles size provides a clue to the lifestyle of a sea cucumber. Stubby finger-like tentacles indicate deposit feeders and finely branched tentacles belong to suspension feeders. Also, look at the locations of ambulacral rows. DRAW what you see (refer to **Fig. 11.12**).

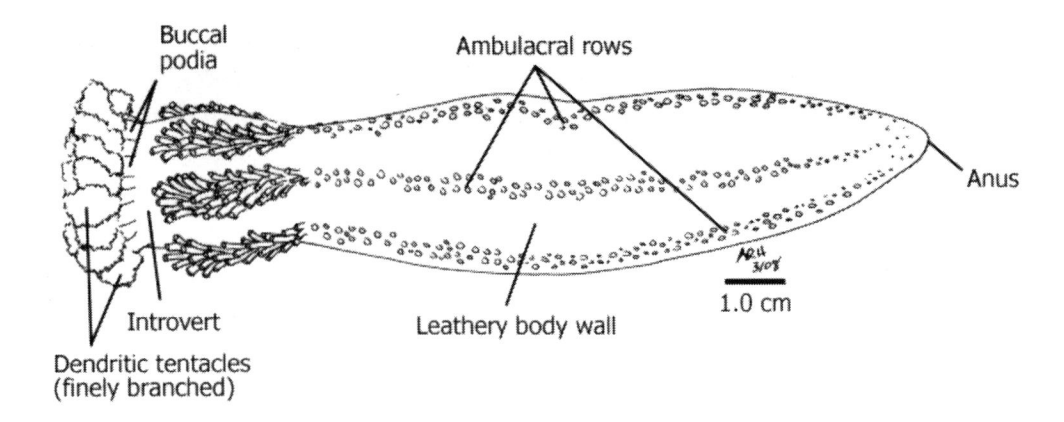

Figure 11.12. External anatomy of the sea cucumber *Cucumaria miniata*. (Image: ARH)

Echinoderm development

Echinoderm development has been studied extensively because they, like humans, are deuterostomes. We have learned a great deal about human development by studying the early embryonic development of echinoderms.

Tasks – Echinoderm developmental stages

1) Use a compound microscope to examine prepared slides of developmental stages of echinoderms.
2) DRAW and label all developmental stages available to you. Refer to **Fig. 11.13** (on the next page) to help you identify the developmental stages as well as their anatomy.

Group Questions

1) What evidence is there that echinoderms have bilateral bodies?
2) Explain why YOU think radial symmetry is a favorable adaptation for slow moving, benthic chemosensory hunters like sea stars.
3) Many seastars and sea urchins have pedicellaria. What advantage is there to investing energy to produce these structures of the body wall?

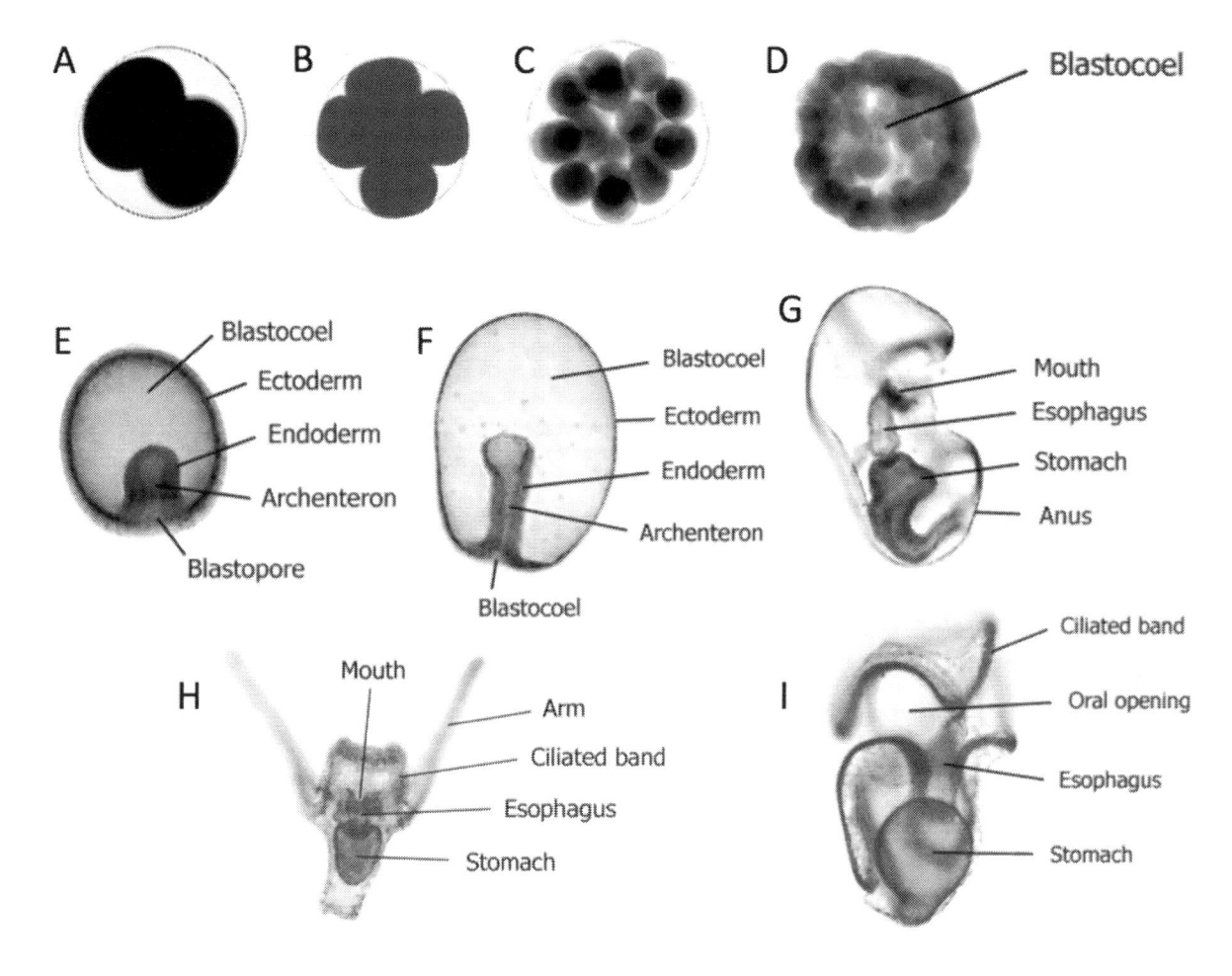

Figure 11.13. Embryonic and larval stages of echinoderms: **A)** 2-cell stage with fertilization membrane, **B)** 4-cell stage with fertilization membrane, **C)** morula stage with fertilization membrane, **D)** early blastula stage, **E)** early gastrula stage, **F)** late gastrula stage, **G)** early bipinnaria larva, **H)** sea urchin two-arm pluteus stage, **I)** sea star bipinnaria larva. (Images: ARH)

Phylum Echinodermata – Glossary

Aboral arm plate (brittle star) – calcified plates that protect the aboral surface of brittle star arms

Aboral longitudinal muscle – muscle that runs along the inner aboral surface of seastar arms, allows seastars to raise their arms

Aboral nerve – nerve that innervates muscles and structures of the aboral body wall in seastar arms

Aboral sinus ring (sea urchin) – aboral connection between the five mesenteries that support the gonads and bears gonopores on the aboral ring

Aboral surface – surface of the body opposite the surface that bears the mouth

Ambulacral groove – depression that runs along the aboral surface of arms of seastars, houses the radial nerve and radial and lateral canals of the water vascular system

Ambulacral plate (echinoid) – calcified plate of the test that bears holes through which podia connect to the rest of the water vascular system

Ambulacral ridge – raised ambulacral structure as viewed from within a seastar arm, houses the radial canal of the water vascular system and radial nerve

Ambulacral rows (sea cucumber) – rows of podia that run along the length of a sea cucumber body, positions of these rows varies between species

Ampulla – the part of tube feet that is housed inside the body, it fills with water when the podium contracts and contraction of muscles surrounding the ampulla extends the podium by hydrostatic pressure, there are also backflow valves that prevent water from being pushed back into the lateral canal from the ampulla

Archenteron – invagination into the blastocoel of a gastrula stage embryo, this is the rudiment of the gut

Aricula of Aristotle's Lantern – extensions of calcified plates of the test that support Aristotle's Lantern and provides points of insertion for jaw retractor muscles

Aristotle's Lantern – complex chewing mouthpart of some echinoids

Arm spines (brittle star) – spines attached to marginal plates of the arm that protect the arm and podia

Axial organ (sea urchin) – portion of the hemal system associated with the stone canal

Bipinnaria stage – bilaterally symmetrical larval stage of a seastar

Bivium – the two arms of a seastar associated with the madreporite

Blastocoel – fluid-filled space within the blastula stage

Blastopore – opening into the archenteron of a gastrula stage embryo

Blastula – developmental stage with a single layer of undifferentiated cells surrounding a fluid-filled blastocoel

Buccal podia (sea cucumber) – appendages that surround the mouth in sea cucumbers, these are used for feeding and gas exchange

Bursa (brittle star) – invagination of the body wall used for gas exchange and brooding offspring

Bursal slit (brittle star) – opening into a bursa

Cardiac stomach (seastar) – portion of the stomach that can be extended out through the mouth and is the portion of the digestive tract where food is stored temporarily before being moved to the pyloric stomach

Cardiac stomach retractor muscles – these muscles pull the cardiac stomach back into the body through the mouth

Cecum (sea urchin) – saclike extension of the digestive tract located where the esophagus, stomach and siphon diverge

Central disc – structure located in the center of crinoids, asteroids and ophiuroids, bears the mouth and anus and gives rise to the arms

Central plate (brittle star) – skeletal plate located in the center of the aboral surface of the central disc

Closer muscle (of pedicellaria) – muscle used to pinch jaws of pedicellaria together

Cryptic – refers to being physically hidden or camouflaged

Deposit feeding – feeding method where organic material is mopped up off of the substrate

Ectoderm – embryonic tissue layer that gives rise to the epidermis, nervous system and associated structures

Elastic ligament (of seastar pedicellaria) – structure that causes pedicellaria to gape open

Endoderm – embryonic tissue layer that gives rise to the digestive tract and associated structures

Endoskeleton – hard internal skeletal support structure, is covered by a single layer of epidermal cells in echinoderms

Enterocoely – pattern of development where the mesoderm arises from outpocketings of the archenteron

Food grooves (sand dollar) – ciliated grooves that carry food particles to the mouth

Gastrula – developmental stage that contains an archenteron and produces embryonic tissues

Genital plate (echinoid) – set of five calcified plates that surround the periproct, each one bears a gonopore including one that bears the madreporite

Gonopore – opening through which eggs and sperm are released

Hyponeural radial canal –fluid-filled canal of the hyponeural system, located aboral to the radial nerve

Interambulacral plate (echinoid) – calcified plate of the test that does not have holes for podia

Introvert (sea cucumber) – the tentacles and related structures that can be retracted into the body

Jaw plate (brittle star) – five plates that surround the mouth and bear the jaws

Jaw protractor muscle of Aristotle's Lantern – muscle that extends the teeth

Jaw retractor muscle of Aristotle's Lantern – muscle that pulls the teeth back up/in

Larva – developmental stage between hatching and metamorphosis into the juvenile stage

Lateral canals – short canals of the water vascular system that carry water from the radial canals to the tube feet

Ligament of Aristotle's Lantern – connective tissue that supports and connects calcified structures of the lantern to each other

Madreporite – a sieve plate through which water is pulled into the water vascular system

Marginal plates (brittle star) – calcified plates that protect the lateral sides of the arm and bear arm spines

Mesentery – mesodermal tissue that attaches an organ to a body wall

Mesoderm – embryonic tissue layer that gives rise to the peritoneum, muscles and other structures between the epidermis and the gut, also lines and protects internal organs and anchors organs to body walls

Morula – developmental stage where the embryo is a solid ball of cells

Muscle joining radii of Aristotle's Lantern – circular muscle connecting radii to each other that pulls upward on structures within the ring of pyramids and helps retract the teeth

Ocular plates (urchin) – five small calcified plates located between the genital plates

Oral arm plate (brittle star) – calcified plate that protects the oral surface of the arm

Oral plate (brittle star) – five large plates on the oral surface of the central disc between the arms, one of these bears the madreporite

Oral surface – surface of the body that bears the mouth

Papula (dermal gill) – thin-walled extension of the body wall of a seastar, used for gas exchange

Pedicellaria – tiny pinching organs of seastars and sea urchins, used to keep the body wall free of epibionts (setting organisms) and for food capture by some seastars

Pentaradial symmetry – body plan where the body produces five sets of structures that radiate from a central axis

Periproct (urchin) – soft tissue layer bearing many tiny calcified plates and surrounding the anus

Peristomial membrane – soft tissue that surrounds the mouth

Perivisceral coelom – fluid-filled space that houses organs of the body and is lined by mesodermal peritoneum

Petaloid – flower-shaped structure on the aboral surface of sand dollars, the slits of the petaloid are dorsal ambulacra or openings through which dermal gills are extended

Pluteus – bilaterally symmetrical developmental stage of sea urchins, there are 2-arm, 4-arm, 6-arm and 8-arm pluteus stages

Podia – portion of the tube foot that extends beyond the body wall

Pyloric stomach (seastar) – portion of the stomach that moves food from the cardiac stomach to the digestive cecae for digestion and absorption

Pyramid of Aristotle's Lantern – calcified structure that houses a tooth and provides it with a groove in which slides up and down as it is extended and retracted

Radial canal – canal of the water vascular system that carries water from the ring canal supplies water to the lateral canals

Radial cleavage – pattern of early embryonic development where four of the cells in the 8-cell stage rest directly on top of the four cells that produced them

Radial nerve – nerve that runs along the floor of the ambulacral groove

Radial shield (brittle star) – sets of aboral plates located at the junction of the central disc and arms, these are visible in some species but appear only as bumps under the epidermis of others

Radius of Aristotle's Lantern – calcified structure that works as a lever that is involved in extending and retracting the teeth

Rectal cecum (seastar) – organ that stores waste material until it is released through the anus

Rectum (seastar) – short tube between the pyloric stomach and anus, bears the rectal cecae

Ring canal – structure of the water vascular system that carries water from the stone canal to the radial canals

Siphon (sea urchin) – a tube attached to the margin of the intestine from the junction of the esophagus and stomach to the junction of the intestine and rectum, it removes excess water from the intestine

Spiral cleavage – pattern of early embryonic development where the four cells produced by the 4-cell stage rest in the depressions between the four cells that gave rise to them instead of directly on top of them

Stellate – star-shaped

Stone canal – structure of the water vascular system that carries water from the madreporite to the ring canal

Suspension feeding – feeding strategy where floating particles are captured and ingested

Test – the rigid endoskeleton of some echinoderms

Tiedemann's bodies – enlarged areas of the ring canal where amoebocytes are reportedly produced that keep the water vascular system free of foreign bodies (immune system)

Trivium - three arms of a seastar that are not associated with the location of the madreporite

Tube feet – portion of the water vascular system used for locomotion, feeding and sensing the environment, may include external podia and internal ampulla or just podia, varies by taxon

Tubercle – knob-shaped structure of a plate of the test that supports a spine in a ball and socket arrangement, also has provides insertion points for spine muscles

Water vascular system – unique hydraulic system of channels, canals and podia found only in echinoderms

Chapter 12: Phylum Hemichordata and Phylum Chordata

Phylum Hemichordata includes about 130 species. These animals were once thought to belong to Phylum Chordata but detailed anatomical and phylogenetic analyses show that they do not and are instead accepted as the sister taxon of Phylum Chordata. Hemichordates are assigned to two classes: Class Enteropneusta, the deposit-feeding acorn worms and Class Pterobranchia, the suspension-feeding pterobranchs. Characteristics of hemichordates are listed in **Table 12.1**.

Table 12.1. Characteristics of Phylum Hemichordata (after Brusca, *et al.* 2016).

- Bilateral deuterostomes with a trimeric body (prosome, mesosome and metasome coelomic spaces)
- Pharyngeal gill openings
- Open circulatory system
- Stomochord
- Short dorsal hollow nerve cord
- Collagenous support structures in the proboscis and gill bars
- Tornaria larva (when larval stage is present)

Phylum Hemichordata "half string" – Taxonomy

Class Enteropneusta "intestine breathing" (covered in this exercise) – acorn worms or tongue worms, about 120 species. Many pairs of gill pores line the branchial region of the digestive tract in these animals and give this class its name. The bodies of acorn worms are divided into three main parts: proboscis, collar and trunk and these correspond to the coelomic spaces referred to in **Table 12.1**. The proboscis houses a hydrostatic skeleton that these worms use to carry out peristalsis as they burrow through soft sediments and move within u-shaped burrows. The proboscis is ciliated and used for deposit feeding. The collar bears the mouth that leads to a linear digestive tract, and the trunk makes up the rest of the body. The trunk has three specialized regions. The branchiogenital region bears gill pores and the gonads, the hepatic region is the region where most of the digestion and nutrient storage occurs and the intestinal or abdominal region carries waste material to the anus where it is released as sediment-rich fecal castings.

Class Pterobranchia "wing gill" (not covered in this exercise) – pterobranchs, about 14 species. These animals bear a pair of branched appendages used for suspension feeding. Pterobranchs are colonial and live in a commonly secreted mass of tubes. The pterobranch body also has three main parts: cephalic shield or proboscis, mesosome and trunk, again these correspond to the coelomic spaces referred to above. The cephalic shield is ciliated and the animal uses it to creep up the inside of a tube to its opening where it can feed. The mesosome bears the mouth and the tentacles. The trunk houses a u-shaped gut and is attached to a muscular stalk that tethers the animal to the rest of the colony and quickly pulls the animal into its protective tube as a defensive strategy.

Class Enteropneusta

Acorn worms are strictly marine and usually live in benthic sand or mud environments where they are deposit feeders. One claim to fame is that enteropneusts can produce brominated and other aromatic compounds, presumably as a chemical defense.

Tasks – Class Enteropneusta

1) External anatomy. Rinse a specimen and immerse it. Yellowish fluid, a brominated substance produced by the worm may diffuse out of your specimen along with lots of mucus. Use a magnifying lens or dissection scope to examine the external anatomy of your specimen. DRAW what you see and use **Fig. 12.1** to help you identify the anatomy of your specimen.

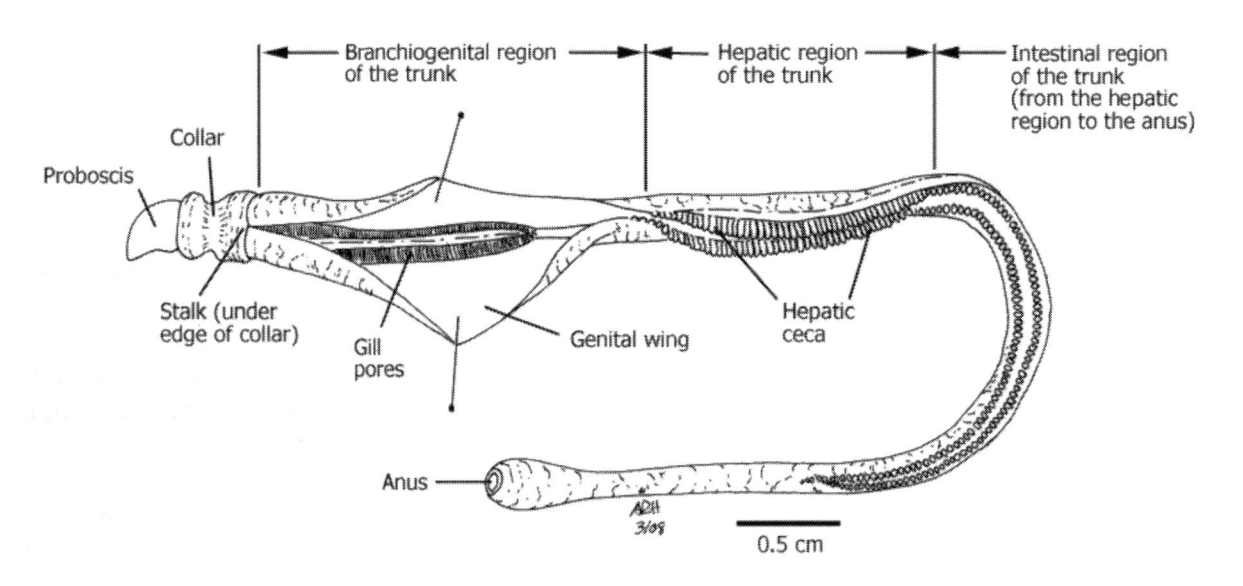

Figure 12.1. The enteropneust acorn worm *Balanoglossus*. The genital wings are pinned back to reveal the gill pores. (Image: ARH)

Phylum Chordata

Phylum Chordata includes three subphyla: Cephalochordata, Urochordata and Vertebrata. The vast majority of chordates including humans belong to Subphylum Vertebrata, animals that produce a backbone. Vertebrates are not included in this laboratory manual but representatives of the other two subphyla are presented. Characteristics of chordates are listed in **Table 12.1**.

Watch this segment from *The Shape of Life* video series to see an introduction to the chordate body plan and to see how animals as diverse as sea squirts and humans can be members of the same phylum: http://shapeoflife.org/video/chordates-we're-all-family

Table 12.1. Characteristics of Phylum Chordata (after Brusca, *et al.*, 2016).

- Bilaterally symmetrical coelomate deuterostomes
- Pharyngeal gill slits/openings
- Dorsal notochord
- Dorsal hollow nerve cord
- Muscular post-anal tail
- Endostyle/thyroid gland

Phylum Chordata "a string" – Taxonomy

Subphylum Cephalochordata "head string" (covered in this exercise) – lancelets, about 30 species. Lancelets are small elongate fish-like animals that lack vertebrae and a cranium. These strictly marine organisms make their living as suspension feeders. They live in shallow water in clean sand habitats where they burrow tail-first into the sediment. Only their heads protrude above the sediment surface so they can pull water into their mouths as they filter-feed. Lancelets are weak swimmers that tend to stay in one place.

Subphylum Urochordata "tail string" (covered in this exercise) – tunicates, salps and appendicularians, about 3,000 species. Urochordates are also strictly marine. They are called urochordates because all the defining characteristics of phylum chordata are typically seen only in the tail-bearing, swimming life stages. Most urochordates make a living as suspension feeders by pulling water through their pharynx and capturing small particles on sheets of mucus secreted by the endostyle.

Subphylum Vertebrata "vertebra" (not covered in this exercise) – animals with a cranium and backbone including fishes, amphibians, reptiles and mammals, about 58,000 species.

Subphylum Cephalochordata

Cephalochordates were once considered to be the sister taxon of the vertebrates but phylogenetic analysis shows clearly that they are not, urochordates are. Cephalochordates are therefore the sister taxon to the clade containing urochordates and vertebrates. When you look at the body plan of these animals it is easy to see why we used to think that these were the ancestral group that gave rise to early jawless fishes.

Tasks – Subphylum Cephalochordata

1) Obtain a preserved specimen. Use a magnifying lens or dissection scope to look for the metameric chevron-shaped muscle blocks that line the flanks of these animals. Also, look for many cream-colored gonads located along the ventral half of the body.
2) Obtain a prepared whole mount slide of a cephalochordate and study its internal anatomy. DRAW what you see. Refer to **Fig. 12.1** to help you identify the anatomy of your specimen. Take particular care to identify the anatomical structures that qualify this animal to be a member of Phylum Chordata. Before you return the prepared slide, take a

close look at the many cup ocelli that are found along the edges of the dorsal hollow nerve cord. Note the orientation of the openings into these cup ocelli.

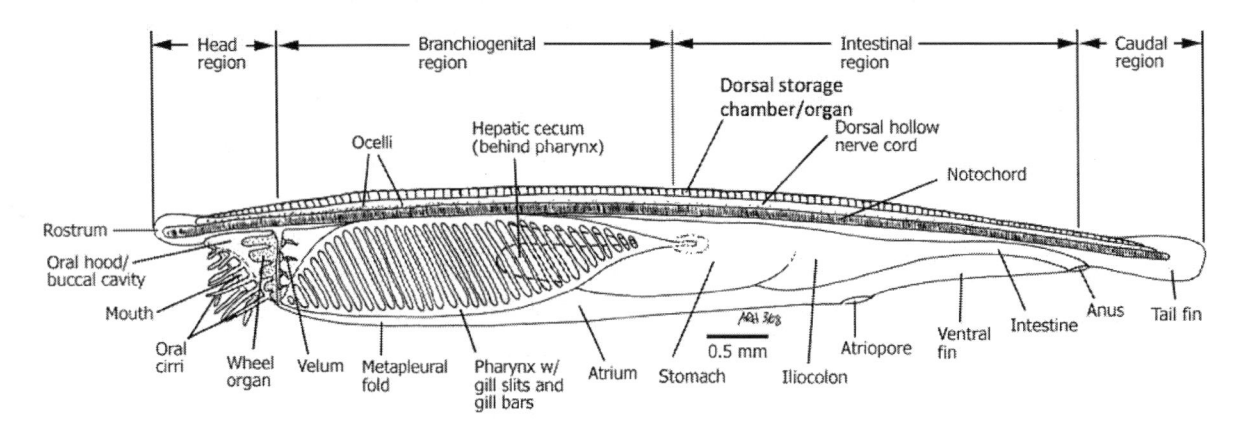

Figure 12.1. Lateral view of the cephalochordate *Branchiostoma*. (Image: ARH)

3) Obtain a prepared cross-section slide through the pharyngeal region of a cephalochordate. Use a compound scope to study the slide. DRAW your specimen and use **Fig.12.2** to help you identify what you see.

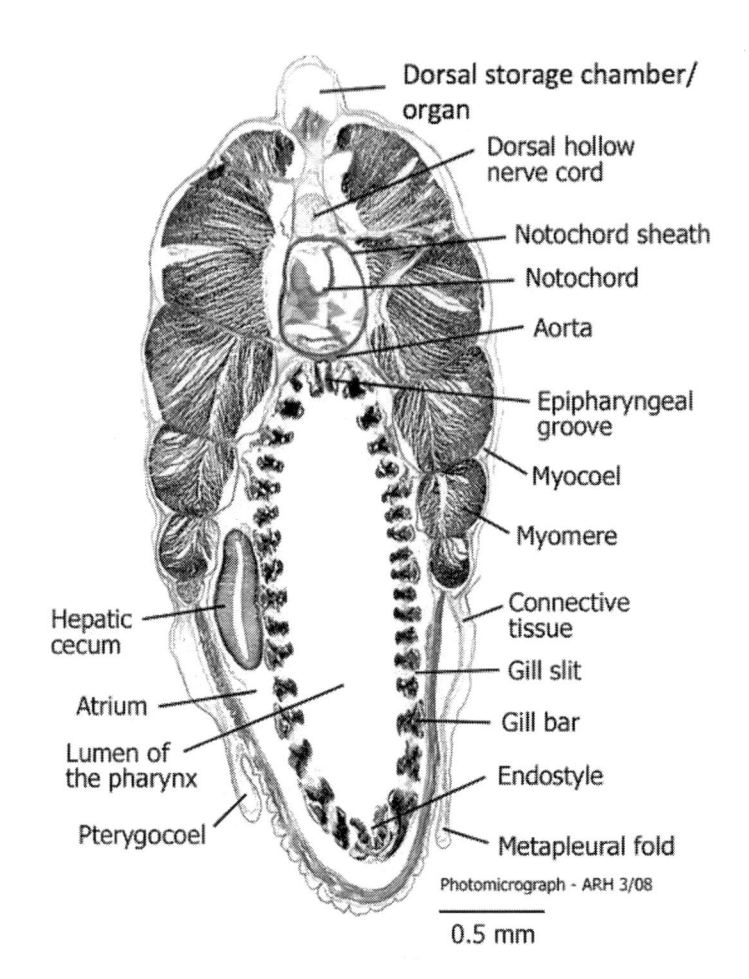

Figure 12.2. Cross-section view through the branchiogenital region of the cephalochordate *Branchiostoma*. (Image: ARH)

Subphylum Urochordata

This subphylum contains four classes: Class Ascidiacea (the sea squirts), Class Thaliacea (salps and doliolarians), Class Appendicularia (larvaceans) and Class Sorberacea (carnivorous urochordates found only in deep water). Of these only Class Ascidiacea is covered in this exercise. This video shows predaceous ascidians, fast forward to the five-minute mark of this video to see what they look like: https://www.youtube.com/watch?v=1PEFmTEF1Sk).

Ascidian tunicates have a sac within a sac body plan. The outer sac includes the tunic and incurrent and excurrent siphons. The composition of the tunic is extremely unusual for any animal, it is made of a substance called tunicin and the primary component of tunicin is cellulose! The inner sac is the pharynx. It is housed within an atrial space. The pharynx has many rows of gill openings called stigmata and is organ used for suspension feeding.

Ascidians do something quite interesting. Every 2-4 minutes they exhibit heartbeat reversal. Yep, the direction of blood flow through the body actually changes direction. The ascidian heart is a contractile vessel without valves so it can do this. There is still debate however over the mechanism that causes this to take place but you can observe heartbeat reversal in the ascidian tunicate *Clavelina* in this short video: https://www.youtube.com/watch?v=Gk-yPmjGxpA

Tasks – Subphylum Urochordata, Class Ascidiacae

1) External anatomy of the solitary tunicate *Molgula*. This tunicate lives attached to just about any hard substrate and prefers calm waters of harbors and bays around the world. Carefully remove the debris and sediment that is commonly found on the tunic. Examine both lateral views of your specimen. Some internal anatomy can be discerned through the translucent tunic. DRAW what you see and refer to **Fig. 12.3** to help you identify the anatomy of *Molgula*.

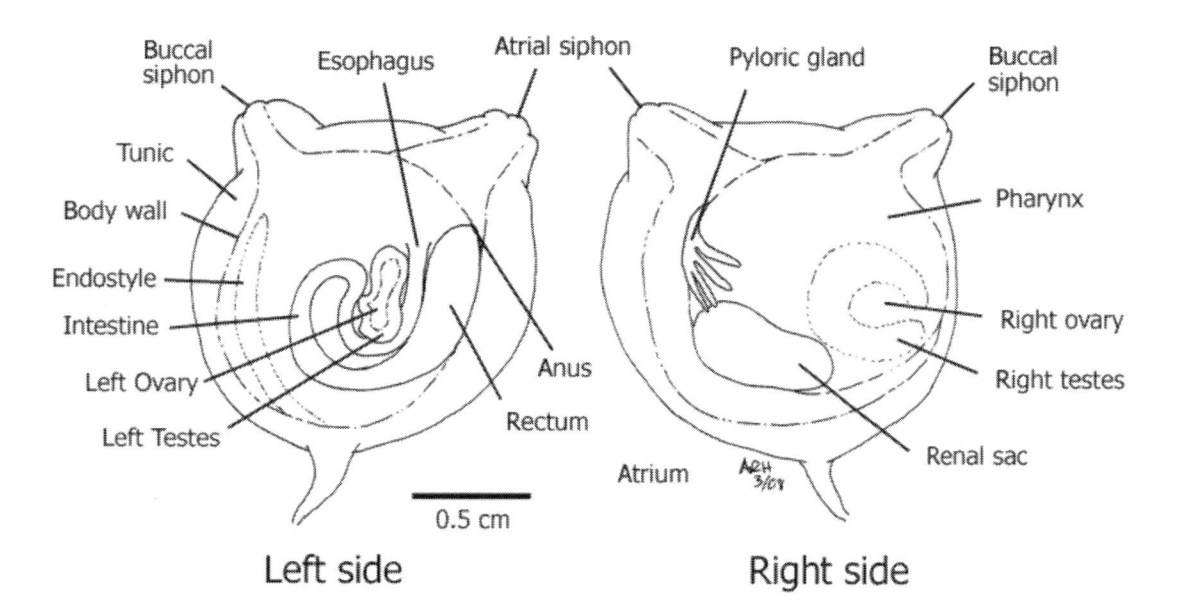

Figure 11.2. External anatomy of *Molgula*, the sea grape. (Images: ARH)

2) Anatomy of the colony-forming tunicate *Ecteinascidia*. This sea squirt lives in tropical waters especially in mangrove communities where they are attached to submerged mangrove roots. This species produces clusters of individuals that are attached to each other via a basal stolon. The tunic of *Ecteinascidia* is largely transparent so its internal anatomy is easily to see. Use a dissection scope or compound scope to study the anatomy of a prepared slide of *Ecteinascidia*. Keep in mind as you observe your specimen that the zooid has been flattened to create the slides you are studying so try to imagine what the specimens would have looked like in their inflated three-dimensional form. DRAW your specimen and use **Fig. 12.4** to help you identify what you see.

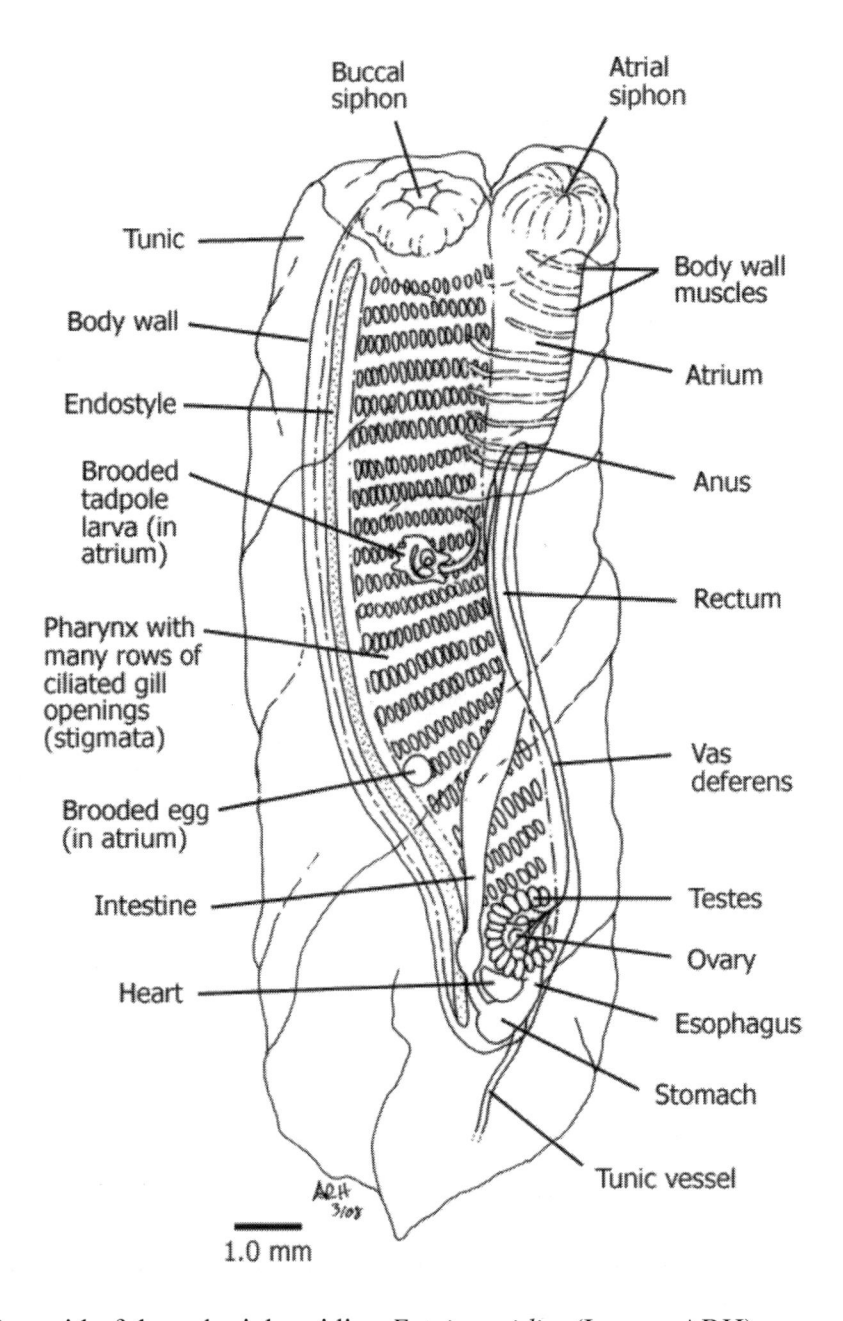

Figure 12.4. A zooid of the colonial ascidian *Ecteinascidia*. (Image: ARH)

3) Ascidian tadpole larva. Ascidians produce a swimming, non-feeding larval stage called the tadpole larva. This larva got its name due to its superficial resemblance to a frog tadpole larva. Use a compound scope to study a prepared slide of a tadpole larva. The tadpole larva is the only life stage of an ascidian when all defining characteristics of Phylum Chordata are present. DRAW your specimen and refer to **Fig. 12.5** to help you identify what you can see.

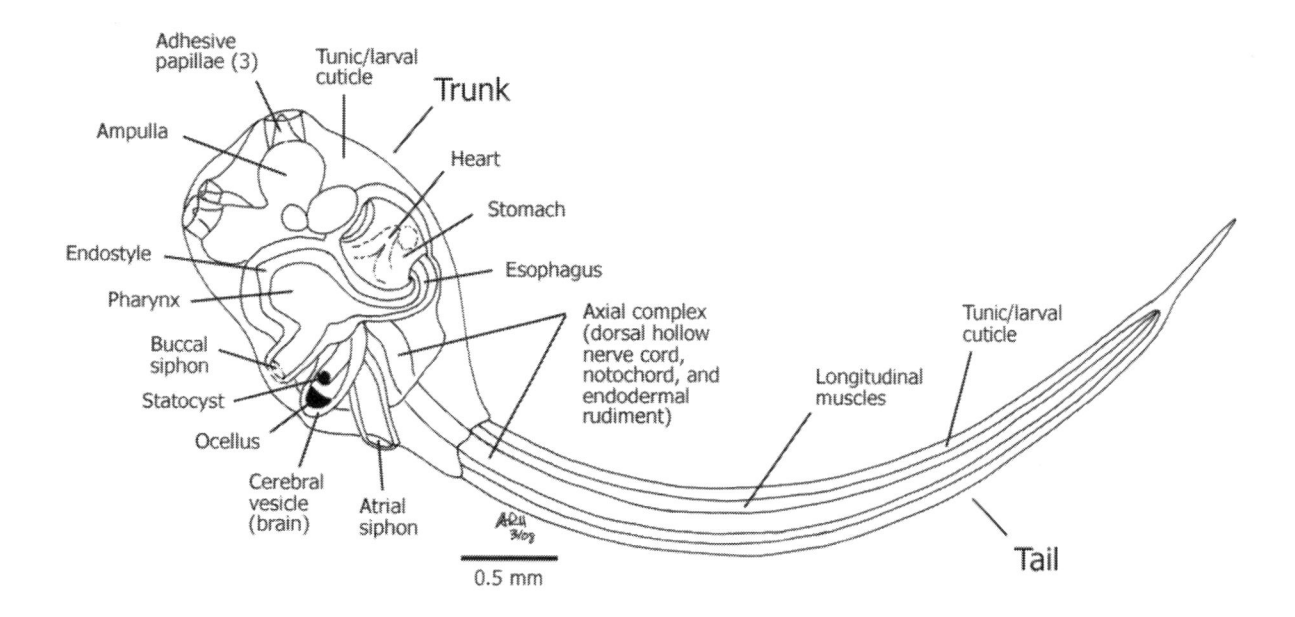

Figure 11.4. Tadpole larva of an ascidian tunicate. (Image: ARH)

Group Questions

1) The body plan of ascidian tunicates is often described as a sac-within-a-sac body plan. How would you describe the body plan of cephalochordates?
2) While you are thinking about it how would you describe the body plan of humans?
3) Explain how animals as different as humans and ascidians are members of the same phylum.

Phylum Hemichordata and Phylum Chordata – Glossary

Adhesive papilla – structure that secretes adhesive material and is used by the tadpole larva to glue itself to the substrate during settlement

Ampulla – bulbous structure that supports the adhesive papillae in the tadpole larva

Aorta – also called the dorsal vessel

Atrial siphon – excurrent siphon of urochordates

Atriopore – water that passes through the pharynx enters the atrium and leaves the body via this excurrent opening in lancelets

Atrium – fluid-filled space surrounding the pharynx of cephalochordates and urochordates

Axial complex – junction of the tail and trunk in tadpole larvae

Branchiogenital region – anterior section of the trunk that bears many gill pores/slits and sac-like gonads in hemichordates

Brooded egg/tadpole – fertilized eggs and larvae develop and are brooded in the urochordate atrium until they are ready to be released

Buccal siphon – incurrent siphon of urochordates

Caudal fin – the tail fin

Cerebral vesicle – this is the brain of tadpole larvae, contains an ocellus and a statocyst

Collar – band body region between the proboscis and trunk of enteropneusts, supports the proboscis, bears the mouth and contains mesosome coelomic spaces

Cranium – cartilaginous or bony protective covering of the brain

Cup ocelli – light sensory organs that are able to discern light direction and intensity but do not generate an image

Dorsal hollow nerve cord – tube of ectodermal origin that forms by infolding and gives rise to the brain and nerve cord, differs from nerve cords of other invertebrate phyla that are ventral and solid

Dorsal storage chamber/organ – a series of box-like structures located along the dorsal midline of cephalochordates used for nutrient storage in preparation for gamete production, this is not a fin rudiment as once thought

Endostyle/thyroid gland – glandular organ located along the ventral wall of the pharynx, secretes the hormone thyroxin and in invertebrate chordates also secretes sheets of mucus used for suspension feeding, the endostyle is homologous to the vertebrate thyroid gland

Epipharyngeal groove – ciliated organ located along the dorsal wall of the pharynx, this is where sheets of mucus secreted by the endostyle are coiled into a mucus rope that is continuously swallowed

Esophagus – passageway between the pharynx and stomach

Genital wing – flap-like extensions of the body wall of enteropneusts that cover gill openings and house gonads

Gill bar – skeletal structure that supports the pharynx and keeps gill slits open

Hepatic ceca (of enteropneusts) – outpocketings of the gut in the hepatic region of the trunk, these increase surface area of the gut and is where most digestion and nutrient storage occurs

Hepatic cecum (of cephalochordates) – outpocketing of the gut that extends anteriorly along one side of the pharynx

Hepatic region (of enteropneusts) – middle section of the trunk, bears many hepatic ceca that are brownish or greenish in color and give this section of the trunk its name

Iliocolon – portion of the digestive tract of cephalochordates where most extracellular digestion takes place

Metameric – refers to any serially repeated structure in a body, e.g., body segments, vertebrae, muscle blocks

Metapleural fold – outpocketing of the coelom, used as a nutrient storage organ like the dorsal storage organ, not a fin rudiment as once thought

Myocoel – coelomic sac that is filled with muscle tissue

Myomere – muscle block

Notochord – stiff but flexible dorsal rod that provides structural support and is mesodermal in origin, it has a sheath that houses many disc-shaped elements

Notochord sheath – tube-shaped covering that surrounds the notochord

Ocellus – see "cup ocelli"

Oral cirri – tentacle-like structures located at the opening of the buccal cavity in cephalochordates, these are sensory and keep large particles from entering the pharynx

Oral hood/buccal cavity – partially enclosed space below the rostrum that leads to the pharynx

Pharyngeal gill slit/opening – openings in the pharynx used for respiration and feeding

Pharynx – structure located between the mouth and stomach, is enlarged and used for suspension feeding in invertebrate chordates

Proboscis – ciliated anterior structure of enteropneusts, used for deposit feeding and burrowing and contains the prosome coelomic sac

Pterygocoel – fluid-filled space within a metapleural fold

Pyloric gland – organ of the digestive gland that produces digestive secretions

Renal sac – organ that collects nitrogenous waste from the blood and stores it as urate crystals with apparently no way to get rid of that waste material

Rostrum – anterior portion of the cephalochordate head located above the oral hood

Statocyst – organ used to sense gravity

Stigmata – openings in the pharynx of urochordates

Stomochord – a hollow tube that extends into the proboscis of enteropneust hemichordates, developmentally it is an extension of the embryonic digestive tract

Suspension feeding – feeding by capturing small particles or organisms from the water column

Tadpole larva – swimming larval stage of ascidian tunicates, includes a trunk and tail and is the only life stage where all chordate traits are present in this group

Tail fin – caudal fin

Tornaria larva – larval form unique to hemichordates, it has ciliary bands running completely around the base of the larva as well as a ciliary band that undulates around the upper portion of the larva, is somewhat reminiscent of a seastar bipinnaria larva

Trimeric body – body plan that has prosome, mesosome and metasome coelomic spaces

Tunic – thick cellulose-based protective outer covering of ascidian tunicates

Tunic/Larval cuticle – protective outer covering of the tadpole larva, covers incurrent and excurrent openings until metamorphosis, if this cuticle is not shed during metamorphosis the larva dies

Tunic vessel – vessel(s) that carries blood to living cells of the tunic

Tunicin – cellulose-based material secreted by ascidian tunicates, primary component of the tunic

Vas deferens – tube that carries sperm from the testes to the atrium in ascidians

Velum – transverse sheet of tissue bearing an opening that joins the buccal cavity and the pharynx and regulates what enters the pharynx

Wheel organ – organ located in the buccal cavity, has ciliated stubby finger-like projections and its cilia together with cilia lining the pharynx pulls water into the pharynx

Supplemental Material: The Laboratory Notebook

This information applies to you if you are a student in my course (Holyoak).

You are required to produce a laboratory notebook, it should contain a complete record of the work you do in the laboratory portion of the course. The main goal of having you do this is to help you learn to work and think as a scientist. I cannot overstate the importance of keeping accurate records and developing a well-formatted notebook, especially if you plan to pursue a career in the sciences. The ability to do these things will yield great dividends when you begin carrying out your own research and you begin recording your own observations and data in research notebooks. This section includes ideas about how you can develop a useful laboratory notebook.

You are required to have a 9" x 12" spiral bound hard-covered artist's sketchbook. The cover protects your work throughout the course and beyond. The spiral binding allows the notebook to lie flat on a table or desktop while you work. The paper is also acid-free so it will not yellow or become brittle over the years. The spiral binding is particularly handy while you are observing something and drawing it or jotting down notes about it at the same time. You also need a couple of 3H pencils. The lead in regular #2 (HB) pencils is too soft and will smear over time. 3H pencils produce a nice line and have lead that is hard enough that it doesn't smear easily. You also need a ruler with metric divisions.

Also, you should not make your laboratory notebook a re-copy notebook where you do work on loose sheets of paper or in another notebook and then recopy your work into the notebook later in an effort to make it prettier. The notebook is designed to be a working tool that you develop throughout the entire course. I frankly don't care that much about how pretty your notebook is as long as it is complete, well organized and contains a record of the work you did in lab.

Notebook formatting guidelines

1. Contact information: You are required to write your name and contact information inside of the front cover of the notebook. You will put a lot of work into this notebook and including contact information in it will increase the chances of it being returned if it gets lost or stolen.
2. Microscope ocular micrometer calibrations or aperture field width measurements: You are required to write the ocular micrometer measurements inside the front cover of the notebook for the microscopes you use. If microscopes don't have ocular micrometers you should still write field width measurements inside the front cover. You will refer to these measurements constantly throughout the course because you need this information to generate scale bars to accompany your lab drawings.
3. Table of Contents: You are required to leave the first two or three pages of the notebook blank when you make your first entry. These pages are for the Table of Contents. A well-designed Table of Contents is essential to quickly and easily finding what you are looking for. Each entry in the Table of Contents should include a page title, page number and the date the work was done.
4. Page formatting for the notebook: You are required to write a page number, date and page title at the top of every page of the notebook. A page title is just a brief description

of what's on the page. It can be as simple as "Mollusc Lab #1" or a description of whatever is on the page.
5. Page layout: You should follow this prescribed layout for each page that contains a drawing. Divide the page into three sections. A large top panel will contain the drawing and two smaller panels at the bottom are for the observations and questions you generate in relation to that drawing. See guidelines for observations and questions later in this section.

Laboratory Drawings

You will produce many drawings during this course. Why do I have you do this? Most people roam the planet looking at lots of things but noticing and seeing very little. For example, they look around and see "a house", "a tree", "a person", "a dog", etc., but they don't go any farther than that. Artists take this one step farther - they look for additional information. They look at a tree and they see patterns of light and dark, the texture of the bark, the angles and patterns of branching, the sizes and shapes of leaves, etc. Scientists take one more step than the artist and look for even more information. Scientists strive to see what they are looking at and try to describe and explain what they see. They do this by making observations and asking questions. A scientist may ask, "Is the branching pattern of this tree adaptive, and if so, how?" Scientists then collect observations and use their data to try to answer their questions. Producing laboratory drawings should help you to develop and hone the observational skills of an artist and the curiosity of a scientist. Your work pattern during lab should be to look, see, draw and then ask questions. As you engage in this pattern of observing, drawing and asking you should develop observational and question-asking skills used by scientists.

Drawing also forces you to look for detail. It has been said that a pencil is a great aid to observation. Most of the drawings you will produce are assigned in the laboratory exercises but in order to gain full benefit from the lab experience you should push yourself to produce additional drawings of what you see. Remember, the goal of producing lab drawings is to help you observe what you are looking at as well as to provide a record of what you did during lab.

Once more, lab drawings are working drawings and as such need not be artistically beautiful. Working drawings help you recall what you saw and did in lab. To that end lab drawings should represent what YOU saw, not what is shown in drawings or photographs in the lab exercises or in the textbook. Draw what YOU see, don't just replicate a drawing from the lab manual. The entries in your lab notebook should provide you with a good enough record of your experiences so that if the occasion requires you can turn to your lab notebook for review rather than cutting up or observing another specimen.

Guidelines for lab drawings

1. Draw big! Fill at least half of the page with one drawing and in most cases have only one drawing per page. This is important because the larger a drawing is the more detail you are likely to include in it. Drawing large also forces your eye to look for detail. Try to include everything you see in your drawings even though you don't know what everything is. This will in turn lead you to asking better questions.
2. Each drawing needs a scale bar. Instructions for generating scale bars were given in Chapter 1 (pp. 12-13). Refer to that section for questions about scale bars.

3. Begin each drawing by including the most obvious things and then add details as you make further observations and see more.
4. Don't hesitate to include more than one drawing from a single dissection.
5. Always use a 3H pencil. Never use a pen for anything in your notebook.
6. Do not shade or color your drawings. Shading or coloring often obscures detail rather than enhances it. Instead, describe the things you see via accompanying observations.
7. Use drawing aids as you see fit, i.e., rulers, circle templates, compasses, calipers, etc. Do not expect drawing aids to be provided. I have observed that some students have taken to using their cell phones to photograph specimens but photographs cannot replace hand-produced lab drawings for developing observational skills. Plus, photographs are not an acceptable substitute for required laboratory drawings when it is time to grade lab notebooks.

Guidelines for Observations and Questions

You are required to label all the structures that you can on each drawing you make. You are also required to generate at least three observations and at least three questions for each drawing/entry you make in your notebook. Observations can clarify what you see as you develop a drawing or they can be thoughts you have about your specimen as you study it. Questions can be about any aspect of the specimen you are observing. They will, I hope, be insightful and often lead you to additional questions and observations.

If you are not sure where to start when it comes to posing questions I recommend that you think about questions that start with "I wonder (how/if/what/why/when)..." Also, any question you would ask your neighbor or the instructor about your specimen could and perhaps should be jotted down. And, you can certainly have more than three questions per entry.

I recommend having designated spaces on each page for a drawing, for observations and for questions. One approach that I have found to be particularly helpful is to have you draw a line 6 to 8 cm above the bottom of a page and then divide the area below the line in half with a vertical line. Use the large area above the line for your drawing, one of the smaller areas below the line for your observations and the other one for your questions – label which section is for questions and which is for observations, this makes laboratory notebook grading much easier.

Reference Material

Bogtish, BJ, and TC Cheng. 1998. Human Parasitology, 2nd Ed. San Diego, CA: Academic Press. 484 pp.

Borror, DJ. 1960. Dictionary of Word Roots and Combining Forms. Mayfield Publ. Co., Palo Alto, CA. 134pp.

Brusca, RC, W Moore, and SM Shuster. 2016. Invertebrates, 3rd Ed. Sunderland, MA: Sinauer Assoc. 1104 pp.

Bullough, WS. 1958. Practical Invertebrate Anatomy. 2nd Ed. London, England: MacMillan. 483 pp.

Cavalier-Smith, T. 2010. Biology Letters. 6: 342-345.

CDC, Centers for Disease Control, http://www.cdc.gov, as indicated in the body of the text.

Endangered Species International, 2012. http://www.endangeredspeciesinternational.org/coralreefs.html

Fox, R. 2001. Invertebrate Anatomy OnLine. *Molgula*. Lander University. http://webs.lander.edu/rsfox/invertebrates/molgula.html

Fox, R. 2001. Invertebrate Anatomy OnLine. *Nacreus americanus*. Lander University. http://lanwebs.lander.edu/faculty/rsfox/invertebrates/narceus.html

Fox, R. 2001. Invertebrate Anatomy OnLine. *Scutigera coeloptrata*. Lander University. http://lanwebs.lander.edu/faculty/rsfox/invertebrates/scutigera.html

Fox, R. 2005. Invertebrate Anatomy OnLine. *Ecteinascidia*. Lander University. http://webs.lander.edu/rsfox/invertebrates/ecteinascidia.html

Fox, R. 2007. Invertebrate Anatomy OnLine. *Nereis viriens*. Lander University. http://webs.lander.edu/rsfox/invertebrates/nereis.html

Harmer, SF, and AE Shipley, eds. 1910. Worms, Rotifers, and Polyzoa. The Cambridge Natural History, Vol. II. London, Macmillan & Co. Ltd. 560 pp.

Harmer, SF, and AE Shipley, eds. 1920. Crustacea and Arachnids. The Cambridge Natural History, Vol. IV. London, Macmillan & Co. Ltd. 566 pp.

Harmer, SF, and AE Shipley, eds. 1922. Peripatus, Myriapods, Insects Part I. The Cambridge Natural History, Vol. V. London, Macmillan & Co. Ltd. 584 pp.

Meinkoth, NA. 1981. The Audubon Society Field Guide to North American Seashore Creatures. Alfred A. Knopf, Inc. 799 pp.

Pechenik, JA. 2015. Biology of the Invertebrates. 7th ed. New York, NY: McGraw-Hill Education. 606 pp.

Raymondo, DM. Practical Parasitology. Scientific Image and Consulting, http://www.practicalscience.com/introtick.html

Romoser, WS, and JG Stoffolano, Jr. 1998. The Science of Entomology. 4th ed. Boston, MA: WCB McGraw-Hill. 605 pp.

Ruppert, EE, RS Fox, and RD Barnes. 2004. Invertebrate Zoology: A Functional Evolutionary Approach. 7th ed. Belmont, CA: Thomson Brooks/Cole. 963 pp.

Struck, *et al.* 2007. Annelid phylogeny and the status of Sipuncula and Echiura. BMC Evolutionary Biology 7: 57-68.

Struck, *et al.* 2011. Phylogenomic analyses unravel annelid evolution. Nature 471: 95-98.

Wallace, RL, and WK Taylor. 2002. Invertebrate Zoology: A Laboratory Manual. 6th ed. Upper Saddle, NJ: Prentice Hall. 356 pp.

Wikipedia (various pages, checking mainly taxonomies) https://www.wikipedia.org.

Index

Acorn worm (see Enteropneusta)
African sleeping sickness (see *Trypanosoma*)
Amoeba proteus (*Chaos diffluens*) 22
Amoebozoa, supergroup 21-22
Animalia, Kingdom 40
Annelida, characteristics 108
Annelida, phylum 108-120
Annelida, terms 118-120
Anthozoa, subphylum and class 51-54
Apicomplexa, phylum 25-27
Aplacophora, class 81
Appendicularia, class 201
Arachnida, subclass 144, 149-153
Aristotle's lantern 179, 187-188
Arrowhead flatworm 68 (and see Tricladida)
Arthropoda, characteristics 144
Arthropoda, phylum 18, 19, 121, 140, 144-177
Arthropoda, terms 171-177
Ascaris lumbricoides 126-130
Ascidiacea, class 201-203
Asconoid sponge body plan 41, 43-45
Asterias 180-182
Asteroidea, class 179-182
Asterozoa, subphylum 178
Aurelia 55-56
Balanoglossus 198
Balantidium coli 28
Basket star (see Ophiuroidea)
Blattoidea, order 169
Beard worm (see Siboglinidae)
Bee (see Hymenoptera)
Beetle (see Coleoptera)
Bilateria, clade 67
Bipinnaria larva 191
Bivalvia, class 82, 97-101
Blastula stage 191
Blue crab (see *Callinectes*)
Bonellia viridis (see Echiuridae)
Brachiopoda, characteristics 121
Brachiopoda, phylum 121-125
Brachiopoda, terms 137
Branchiostoma 200

Brittle star (see Ophiuroidea)
Bryozoa, phylum 121
Butterfly (see Lepidoptera)
Caddis fly (see Trichoptera)
Calcarea, class 40, 43, 45
Cambarus 158-161
Capitulum 150-151
Caudofoveata, class 81
Caenorhabditis elegans 125
Callinectes 162-164
Centipede (see Chilopoda)
Cephalopoda, class 83, 90-97
Ceratium 25
Cercaria (*Clonorchis*) 73-74
Cestoda, cohort 69, 75-77
Chagas disease 35
Chelicerata, characteristics 146
Chelicerata, subphylum 144, 146-153
Cephalochordata, subphylum 198-200
Chilopoda, class 154-155
Chinese liver fluke (see *Clonorchis*)
Chiton (see Polyplacophora)
Choanoflagellata, phylum 36-37
Chordata, characteristics 199
Chordata, phylum 18-19, 178, 197-206
Chordata, terms 203-206
Chromalveolata, supergroup 21, 25
Ciliata/Ciliophora, Phylum 28-31
Cladistics, terms 14
Cladogram, how to make 15-17
Clams (see Bivalvia)
Clitellata, family 109, 114-117
Clonorchis (*Opisthorchis*) *sinensis* 72-74
Cnidaria, phylum 50-61
Cnidaria, characteristics 50
Cnidaria, terms 63-66
Cockroach (see Blattoidea)
Coleoptera, order 170
Comb jelly (see Ctenophora)
Coral (see also Anthozoa) 54
Crustacea, characteristics 157
Crustacea, subphylum 58, 144-145, 157-165
Crayfish 158-162
Crinoidea, class 178

Crinozoa, subphylum 178
Cryptochiton stelleri 83-86
Ctenophora, characteristics 62
Ctenophora, phylum 62
Ctenophora, terms 66
Cubozoa, class 51
Cucumaria miniata 190
Daphnia 157-158
Demospongia, class 41
Dendraster excentricus 189
Dermacentor andersoni 151
Dermaptera, order 169
Deuterostomia, clade 178, 190, 197, 199
Didinium 28-29
Digenean flukes (see Trematoda)
Dinoflagellata, Phylum 25
Diplomonadida, phylum 33-34
Diplopoda, class 155-156
Diptera, order 170
Dirofilaria immitis 136
Doliolarian (see Thaliacea)
Dracunculus medinensis 135
Dragonfly (see Odonata)
Dugesia 70
Earthworm (see Clitellata)
Earwig (see Dermaptera)
Ecdysis 67, 121, 125, 140, 141, 144
Ecdysozoa, clade 67, 121, 125-136, 178
Echinodermata, characteristics 178
Echinodermata, phylum 18-19, 178-196
Echinodermata, terms 191-196
Echinoidea, class 179, 183-189
Echinozoa, subphylum 179
Echiuridae, family 109
Ecteinascidia 202
Ectoproct (see Bryozoa)
Eimeria 25-26
Elephantiasis (see *Wuchereria*)
Entamoeba (*Endamoeba*) *hystolytica* 22-24
Enterobius vermicularis 134
Enteropneusta, class 197-198
Ephemeroptera, order 168
Ephyra 54
Errantia, subclass 109, 111-113
Euchelicerata, class 148-153
Euglena 34-35

Euplectella 41-43
Euglenida, phylum 34-35
Eukarya, introduction to 21-39
Eukarya, characteristics 21
Eukarya, terms 37-39
Excavata, supergroup 21, 33
Feather star (see Crinoidea)
Five-Kingdom Model 21
Flea (see Siphonaptera)
Flukes (see Monogenea and Trematoda)
Foraminferans 31-32
Gastropoda, class 82, 87-90
Gastrula stage 191
Gemmule 47, 48
Giardia lamblia (*intestinalis*) 33-34
Glass sponges (see Hexactinellida)
Glochidia 101
Gordian worm (see Nematomorpha and Nematoida)
Granuloreticulosa, Phylum 31-32
Grantia 45-46
Grasshopper (see Orthoptera)
Guinea fire worm (see *Dracunculus*)
Gumboot chiton (see *Cryptochiton*)
Heartworm (see *Dirofilaria*)
Helix 87-90
Hemichordata, characteristics 197
Hemichordata, phylum 178, 197-198
Hemichordata, terms 203-206
Hemiptera, order 169
Hexactinellida, class 41-43
Hexapoda, characteristics 166
Hexapoda, subphylum 145, 165-170
Hirudo 116-117
Holothuroidea, class 179, 189-190
Homoscleromorpha, class 41
Horsefly (see Diptera)
Horsehair worm (see Nematomorpha and Nematoida)
Horseshoe crab (see *Limulus*)
Hydra 51, 58-60
Hydroid or hydrozoan (see Hydrozoa)
Hydrozoa, class 51, 57-61
Hymenoptera, order 170
Insects (see Hexapoda)
Isophyllastrea rigidia 54

Jellyfish, stalked (see Staurozoa)
Jellyfish, box (see Cubozoa)
Jellyfish, true (see Scyphozoa)
Kinetoplastida, phylum 35-36
Kinorhyncha, phylum 121
Lamp shell (see Brachiopoda)
Lancelet (see Cephalochordata)
Larvacean (see Appendicularia)
Leech (see Clitellata)
Lepidoptera, order 170
Leucosolenia 44-45
Leuconoid sponge body plan 46-48
Limpets (see Gastropoda)
Limulus polyphemus 148-149
Lingula 122-124
Lophophorata, clade 121
Loricifera, phylum 121
Lophophore 121-124
Louse (see Psocodea)
Lumbricus 114-116
Malaria (see *Plasmodium*)
Mantodea, order 169
Mayfly (see Ephemeroptera)
Medusozoa, subphylum 51, 55-61
Megalops larva 165
Metacercaria (*Clonorchis*) 73-74
Metazoa, clade 40
Metridium 52-54
Microfilariae (*Wuchereria* and *Dirofilaria*) 133, 136
Microscope, Compound 6-7
Microscope, Dissection 8-9
Microscopy, Rules 9-11
Millipede (see Diplopoda)
Miricidium (*Clonorchis*) 73
Molgula 201
Mollusca, characteristics 81
Mollusca, phylum 81-107
Mollusca, terms 102-107
Monogenea, cohort 69, 71-72
Monoplacophora, class 81-82
Morula stage 191
Moss piglet (see Tardigrada)
Mud dragon (see Kinorhyncha)
Mussels (see Bivalvia)
Myriapoda, characteristics 153

Myriapoda, subphylum 144-145, 153-156
Myxozoa, subphylum 52
Nauplius larva 157, 164-165
Nematoda, characteristics 125
Nematoda, phylum 18, 19, 121, 125-136
Nematoda, terms 137-139
Nematoida, clade 121
Nematomorpha, phylum 121
Neodermata, infraclass 69, 70-77
Nereididae, family 109, 111-113
Nereis 111-113
Nymphon brevirostre 147
Obelia 60-61
Octopus (see Cephalopoda)
Ocular micrometer, how to calibrate 13
Odonata, order 168
Oncosphere (*Taenia*) 75
Onychophora, characteristics 142
Onychophora, phylum 121, 140, 142-143
Onychophora, terms 171-177
Ophiuroidea, class 179, 183-184
Opisthokonta 21, 36
Orthoptera, order 169
Panarthropoda, clade 121, 140-177
Parabasalida, phylum 33
Paramecium 28-29
Parapodia 109, 111-112
Peanut worm (see Sipuncula)
Pedicellaria 180, 186
Pelomyxa (*Chaos*) *carolinensis* 22, 23
Pentatrichomonas (Trichomonas) hominis 33
Peripatus 143
Phasmida, order 169
Phoronida, phylum 121
Phylogenetic analysis (see Cladistics)
Physalia 57
Pinworm (see *Enterobius*)
Planarian (see Tricladida)
Planula 50, 51, 56, 58
Plasmodium 26-27
Platyhelminthes, characteristics 67
Platyhelminthes, phylum 67-80
Platyhelminthes, terms 77-80
Plecoptera, order 168
Pleurobrachia 62

Pluteus larva 191
Polychaete (see Nereididae and *Nereis*)
Polyplacophora 82, 83-86
Polycladida, order (polyclad flatworms) 68
Porifera, phylum 40-49
Porifera, characteristics 40
Porifera, terms 48-49
Porifera, characteristics 40
Portuguese man o' war (see *Physalia*)
Psocodea, order 169
Praying mantis (see Mantodea)
Priapulid or penis worm (see Priapulida)
Priapulida, phylum 121
Proglottid 76-77
Protostomia 178
Pterobranchia 179
Pycnogonida, class 147
Radiolaria, Phylum 32
Rhizaria, supergroup 21, 31
Redia (*Clonorchis*) 73-74
Rhipicephalus (*Boophilus*) *annulatus* 150
Riftia (see Siboglinidae)
Romalea guttata 166-168
Rough star coral (see *Isophyllastrea*)
Roundworm (see Nematoda)
Salp (see Thaliacea)
Sand dollar (see Echinoidea)
Scale Bar, how to make 12-13
Scalidomorpha, clade 121
Scaphopoda, class 82-83
Scolex 76
Scolopendra 154-155
Scorpion 151-152
Scypha 45-46
Scyphistoma 54
Scyphozoa, class 51, 55-56
Sea anemone (see Anthozoa)
Sea cucumber (see Holothuroidea)
Sea gooseberry (see Ctenophora)
Sea lily (see Crinoidea)
Sea spider (see Pycnogonida)
Sea squirt (see Ascidiacea)
Sea urchin (see Echinoidea)
Seastar (see Asteroidea)
Sedentaria, subclass 109, 114-117
Siboglinidae, family 109

Silverfish (see Thysanura)
Siphonaptera, order 170
Sipuncula, family 108, 110
Sipunculus nudis (see Sipuncula)
Slugs (see Gastropoda)
Snail (see Gastropoda)
Solenogastres, class 81
Sorberacea, class 201
Spicule worms (see Aplacophora, Caudofoveata and Solenogastres)
Spider 152-153
Spiralia, clade 67, 121, 178
Spirobolus 156
Spirostomum 30
Sponges (see Porifera)
Spongicola 41, 43
Spongilla 47-48
Spoon worm (see Echiuridae)
Sporocyst (*Clonorchis*) 73
Squid (see Cephalopoda)
Staurozoa, class 51
Stentor 30-31
Stonefly (see Plecoptera)
Strobila 54
Strongylocentrotus 185-188
Syconoid sponge body plan 45-46
Taenia saginata 75-77
Tadpole larva, ascidian 203
Tapeworms (see Cestoda)
Tarantula (see spider)
Tardigrada, characteristics 140
Tardigrada, phylum 121, 140-142
Tardigrada, terms 171-177
Terebratalia and *Terebratulina* 122, 124-125
Thaliacea, class 201
Three Domain Model 21
Thysanura, order 168
Tick (see Arachnida)
Trematoda, cohort 69, 72-74
Trichinella spiralis 131-132
Trichomonads 33
Tricladida, order (triclad flatworms) 68-70
Trichoptera, order 170
Trilobite 145-146
Trilobite larva 148

Trilobitomorpha, characteristics 145
Trilobitomorpha/Trilobita, subphylum 144-146
True bug (see Hemiptera)
Trypanosoma 35-36
Tsetse fly 36
Tubatrix aceti 125
Tunicate (see Urochordata)
Tusk shell (see Scaphopoda)
Unio 97-101
Urechis caupo (see Echiuridae)
Urochordata, subphylum 198, 201-203
Velvet worm (see Onychophora)
Vertebrata, subphylum 198-199
Vinegar eel (see *Tubatrix)*
Vorticella 31
Walking stick (see Phasmida)
Water bear (see Tardigrada)
Water flea (see *Daphnia)*
Wet-mount slide, how to make 11
Wuchereria bancrofti 133
Xiphosura, order 148
Zoea larva 165

Alan Holyoak earned a PhD in Biology at the University of California, Santa Cruz where he was a student of Todd Newberry and focused on the biology marine invertebrate animals. He has taught invertebrate zoology for many years, first at Manchester University, Indiana (1992-2002), and currently at Brigham Young University-Idaho (2002-present). In addition he has taught field courses or carried out research at the Friday Harbor Laboratories of the University of Washington, The University of Hawaii, Joseph M. Long Marine Laboratory of UC Santa Cruz, Hopkins Marine Station of Stanford University and The Oregon Institute of Marine Biology of the University of Oregon.

Made in the USA
Middletown, DE
04 November 2017